WELFARE HORSE SPORTS

What if horse sports didn't just include welfare, but were built around it?

In this timely and transformative book, equine educator and welfare advocate Lisa Ashton offers an evidence-based approach to the use of horses in sport. She presents a compelling blueprint for integrating welfare within the very heart of equestrian practice.

Through a clear and actionable 'welfare map,' Ashton guides riders, coaches, and trainers who are ready to shift their mindsets and skill sets toward more ethical, sustainable approaches. Rather than treating welfare as an add-on, she introduces the concept of welfare horse sports: a model where equine well-being is foundational, not optional.

Drawing on an intersectional and empowering approach, Ashton emphasises that meaningful change doesn't require a title or a platform. It starts with one equestrian—one coach, one rider, one trainer—bringing their unique voice, influence, and compassion to a shared mission: a better life for sport horses.

Welfare Horse Sports is not just a call to action. It's a roadmap for a more ethical and informed equestrian future. For anyone ready to earn public trust, elevate standards, and future-proof horse sports, this book is your starting point.

WELFARE HORSE SPORTS

A BLUEPRINT FOR POSITIVE CHANGE

Lisa Ashton

CRC Press is an imprint of the
Taylor & Francis Group, an **Informa** business

Designed Cover Image: Tara Hodges

First edition published 2026
by CRC Press
2385 NW Executive Center Drive, Suite 320, Boca Raton, FL 33431

and by CRC Press
4 Park Square, Milton Park, Abingdon, Oxon, OX14 4RN

CRC Press is an imprint of Taylor & Francis Group, LLC

© 2026 Lisa Ashton

Reasonable efforts have been made to publish reliable data and information, but the author and publisher cannot assume responsibility for the validity of all materials or the consequences of their use. The authors and publishers have attempted to trace the copyright holders of all material reproduced in this publication and apologize to copyright holders if permission to publish in this form has not been obtained. If any copyright material has not been acknowledged please write and let us know so we may rectify in any future reprint.

Except as permitted under U.S. Copyright Law, no part of this book may be reprinted, reproduced, transmitted, or utilized in any form by any electronic, mechanical, or other means, now known or hereafter invented, including photocopying, microfilming, and recording, or in any information storage or retrieval system, without written permission from the publishers.

For permission to photocopy or use material electronically from this work, access www.copyright.com or contact the Copyright Clearance Center, Inc. (CCC), 222 Rosewood Drive, Danvers, MA 01923, 978-750-8400. For works that are not available on CCC please contact mpkbookspermissions@tandf.co.uk

For Product Safety Concerns and Information please contact our EU representative GPSR@taylorandfrancis.com. Taylor & Francis Verlag GmbH, Kaufingerstraße 24, 80331 München, Germany.

Trademark notice: Product or corporate names may be trademarks or registered trademarks and are used only for identification and explanation without intent to infringe.

ISBN: 9781041000761 (hbk)
ISBN: 9781041000693 (pbk)
ISBN: 9781003608066 (ebk)

DOI: 10.1201/9781003608066

Typeset in Bembo
by Deanta Global Publishing Services, Chennai, India

DEDICATION

For my incredible parents and dear friend, Jen Brown-Watson.

Welfare Horse Sports is fundamentally about the principle of leaving every horse we interact with better. This ethos of continuous improvement, driven by a commitment to truly understanding horses, mirrors the unwavering support you have each provided throughout my equestrian journey.

To my cherished friend, Jen Brown-Watson, thank you for being an indispensable part of my community. This book underscores the vital role of communities of practice in shaping narratives and modelling welfare. Your shared passion for horses, equitation science, and your willingness to engage in open and honest conversations, even when challenging conventional norms, perfectly embody the spirit of collaboration and shared values that are crucial for cascading positive change for horses. Our friendship has been a psychologically safe space where curiosity and growth could flourish. Your unwavering belief in me has been a constant source of strength, empowering me to go first and become the horse's advocate, even when facing difficult conversations. Just as this book champions the horse's perspective, your support encourages me to speak up and contribute to a future where horse welfare is truly prioritised.

Thank you both for being steadfast companions on this continuing path toward deeper understanding and the pursuit of doing better for all horses. Finally, I extend my enduring gratitude to the horses themselves, who have enriched my life, shaped my thinking, informed my practice, and continually inspired my curiosity and capacity for deeper connection.

With heartfelt gratitude, Lisa

CONTENTS

Foreword by Susan Kjærgård   x
About the Author   xii
Glossary of Key Terms   xiv
How to Use This Book   xvi

1 The shift   1
    Reflexive praxis   14
    Is your horse living 'a good life'?   19
      A good life? Checklist   19
    The three Fs   19
      Friends   19
      Forage   20
      Freedom   21
    Reflexive praxis   24

2 The welfare horse sports mindset   25
    Reflexive praxis   27
    Releasing pressure provides positive experience   29
    Reflexive praxis   29
    Five Domains (2020) and First Principles (ISES)   31
    Reflexive praxis   56

3 Tools for transformation: Frameworks and models — 58
Why trust matters more than facts 59
Principle over method 62
Reflexive praxis 66

4 Leading cultural change: Communities and influence — 68
Leave horses better CoP 69
Observable behaviours: Indices 70
Resiliency building 78
LHB: Leave Horses Better 78
Reflexive praxis 80
The importance of well-being 83
Reflexive praxis 84
Reflexive praxis 85
Welfare reform 89
Reflexive praxis 89
Anthropomorphism and reflexivity 92
Looking backwards: The Clever Hans phenonomen 93
Harnessing cascades 96
Social identity and evidence-based horse training 98
Deep dive 102
Reflexive praxis 105
Check the pulse 105
Reflexive praxis 107
Reflexive praxis 108
Triple well-being: Equine, equestrian, and environment: Leadership 110
Professor Natalie Waran, Chair of the Equine Ethics and Well-being Commission (2022–2024) 111
Going first 112
Reflexive praxis 114
Becoming Despacito 115
Reflexive praxis 115
Deep dive: The equitation science smoothie 116
Living into our values 118
Reflexive praxis 118

Reflexive praxis 118
Deep dive journaling 120
Leaning into values 121
Reflexive praxis 121
Resiliency building (horse) 122
Resiliency building (us) 124
Reflexive praxis 124
Why isn't everyone resilient? 125
Empathy building 127
Empathy misses 128
Reflexive praxis 129
Reflexive praxis 130
Positive reinforcement is not just for horses! 130
Reflexive praxis 131
Leave horses better community of practice 133
Guard-rails for communities 134
Leadership is a behaviour not a title 135
Reflexive praxis 138
Building communities of practice 139
Reflexive praxis 149
Becoming your horse's advocate 150

Appendix I: Lesson plans	153
Appendix II: Resources	168
Index	172

ISES Training Principles Flyer

FOREWORD

Horses and humans share more than the space of the arena—we share the same fundamental needs: to feel safe, competent, and resilient. This, as Lisa Ashton beautifully concludes, is the ultimate goal of working with horses. And, as I came to realize while reading *Welfare Horse Sports: A Blueprint for Positive Change*, it is equally the goal of working on ourselves.

Lisa's blueprint for creating Welfare Horse Sports and always leaving horses better than we found them, reaches far beyond the schooling arena. It invites us to reflect deeply—not just on what we do, but on how we think, feel, and choose when it comes to our horses. Guided by the First Principles of Horse Training (ISES) and the Five Domains Welfare Framework, her words ask us the essential question: "If you knew better, would you do better?"

Through the reflexive questions at the end of each chapter, I found myself pausing—not only to consider my actions with horses, but also my choices as a person. Her examples from outside the equestrian world make it clear: curiosity and self-awareness are the true agents of change. Curiosity has helped me work through obstacles that, at first, felt insurmountable. It reminds me that learning, for both horses and humans, requires the same gentle persistence and courage to try again.

Coming from a background as a traditional GP showjumping rider, my own shift toward a horse-centred, evidence-based approach—built on empathy, science, and respect—has not always been comfortable. Change rarely is. But in Lisa's writing, I felt seen. She gives language to the discomfort, compassion to the process, and structure to the journey. Her work encourages us to build authentic connections—not only with our horses, but with each other as equestrians.

This book reminds us that while horses have no direct voice in our policies or practices, we do. We are their advocates. And advocacy, as Lisa reminds us, begins with courage—the courage to know our values and to live into them, even when it's hard.

Welfare Horse Sports offers not just philosophy, but practical pathways—small, compassionate actions that make a real difference to a horse's quality of life. And in doing so, it leaves not only horses better, but equestrians too.

Reading this book has strengthened my own equestrian voice. It has reminded me that how we make people feel is as important as how we make our horses feel. When we, like our horses, feel seen, heard, and valued—we find connection, community, and the resilience to change.

May this book inspire you, as it did me, to stay curious, to lead with empathy, and to work—for horses and for ourselves—toward a shared sense of safety, competence, and resilience.

Susan Kjærgård
www.blueberryhill.dk

ABOUT THE AUTHOR

Lisa Ashton
Equitation Science Consultant
United Kingdom
MBA PGCE BA (Hons) Equitation
Science International (ESI)
Diploma Pony Club A Test
British Horse Society (BHS)
Stage 4 Senior Coach

Lisa is the founder of Equicoach.Life, a global coaching platform committed to advancing ethical, evidence-based horse training. Recently elected as Education Officer for the International Society for Equitation Science (ISES), Lisa educates coaches, riding schools, and welfare organisations in applying evidence-based equitation (EBE). Her consultancy experience includes delivering long-term equine ethics, animal welfare, and cultural change at Mount St John (MSJ) Stud, Blue Cross, Redwings, and the Horse Trust. Her work embeds the Five Domains (2020) Welfare Framework and the application of the First Principles (ISES) in horse training.

Lisa brings an intersectional, rigorous, evidence-based approach to equestrian education, with experience as a Senior Lecturer leading equine cognition and learning, equine welfare and ethics, and equitation science. She is the co-host of the popular podcast *The Other End of the Reins* with Dr Andrew McLean and winner of the 'Expert in Their Field' StAR Award (as voted by students) at Hartpury University. Lisa's outreach is rooted in 20 years of education in the application of learning theory, dedicated to advocating for ethical, sustainable equitation practices through coaching, scholarship, leading, and managing communities of practice (CoP). Her published work includes book chapters on equine behaviour, welfare, and the role of the horse in equestrian coaching, with peer-reviewed contributions in *Animals* and the *Journal of Veterinary Behaviour*.

GLOSSARY OF KEY TERMS

Shaping

- *Definition*: Shaping is the process of teaching a horse a new behaviour by rewarding small steps that lead toward the desired behaviour. Instead of expecting the horse to perform the full action right away, you reward them for each small progress in the right direction.
- *Example*: If you're teaching a horse to step back, you first reinforce (release pressure) for a shift in weight, then a small step, and gradually more steps until the horse can back up easily on cue.

Self-carriage

- *Definition*: Self-carriage means the horse is able to maintain a specific gait/movement without needing constant signals or pressure from the rider. The horse should keep going or stay still (immobility) until signalled/cued otherwise.
- *Example*: When a horse is in self-carriage, you can ask them to walk and they will keep walking without needing you to keep applying leg pressure/leg cues. Similarly, when you ask the horse to stop, they stay immobile, without needing to repeat rein pressure.

Classical conditioning

- *Definition*: Classical conditioning is when a horse learns to associate one thing (a cue) with another thing (a response). This happens when you repeatedly pair a neutral cue (like a cluck) with a known response (scratch) so the horse starts responding to the cue automatically.
- *Example*: If you cluck with your mouth before you add a scratch, the horse learns at hearing the cluck, the scratch is coming, because they have associated that sound with the delivery of scratching.

Parasympathetic nervous system

- *Definition*: The parasympathetic nervous system is the part of the horse's body that helps them relax and rest. It is often activated when the horse feels safe and calm, leading to relaxation responses like slower breathing, relaxed muscles, and blinking more frequently.
- *Example*: When you scratch a horse in its 'sweet spot,' such as the base of the withers, the horse may blink slowly and lower its head, showing that its parasympathetic nervous system is helping it relax.

Valence (affective state)

- *Definition*: Valence refers to whether a horse's emotional state is positive or negative. A positive valence means the horse is experiencing something enjoyable or pleasant, while a negative valence means the horse may be feeling distress or discomfort.
- *Example of positive valence*: A horse that is clucked at and scratched for standing still at the mounting block will begin to associate the mounting block with a positive experience, experiencing a positive valence at the mounting block (context of repeat scratches).
- *Example of negative valence*: A horse receiving clashing signals from a rider—being asked to stop while also being squeezed from the calves to go—horse may at best be confused, leading to a need to resolve this conflict by doing behaviours such as rearing, bucking, or bolting, as a result of experiencing a negative valence.

HOW TO USE THIS BOOK

As pressure grows to demonstrate horse sports with welfare following a series of critical global incidents and 'immense societal change,' questions are being asked about the welfare of sport horses; do equestrian sports push horses to the point of abuse?

Welfare Horse Sports is a blueprint for equestrians curious or currently undertaking shifts in mindset and skills to future-proof horse sports.

We already have books that explore the application of learning theory in horse training in great depth. This book is not one of them. Instead, *Welfare Horse Sports* is written for equestrians who are ready to update their skills and shift their mindset—reframe their approach in order to enhance welfare outcomes. At its heart, this book is shaped by the final part of Maya Angelou's well-known quote: 'Then, when we know better, we do better.' It looks backwards through the lens of 'now we know better,' not as a value judgement on the ethicality of heritage practices, but as an invitation to critically reflect and move forward. It specifically leans into the intersection of learning theory, animal welfare science, and ethics—not as separate silos, but as interwoven forces guiding a new paradigm in horse–human relationships. Just as restoring an old building requires discerning what to preserve and what to modernise, so too does updating equestrian heritage. Traditional practices can be held up to the test of First Principles (International Society for Equitation Science [ISES]). Those that stand up remain part of the structure. Those that do not can—and should—be updated. This book encourages you to observe all horse–human relationships, past and present, with the same evidence-based lens: 'know better, do better.' That includes revered equestrians of history. We must objectively assess the horse's lived experience—even in the past.

This is not a book comparing today's horses to those 100 or 200 years ago. But if it were, we would begin by identifying their negative experiences, evident in equine facial expressions captured in historical art—paintings, sculptures, and artefacts. These cultural documents reveal a great deal, especially when we notice consistent facial expressions that reflect discomfort, pain, confusion, or fear. Equipment on the horse's head in those times, often glorified as traditional or classical, must also be re-examined through the lens of the horse's physical and mental state.

This is why the notion that classical equitation represents a time when horses were spared negative experiences must be met with critical discernment. That is, the ability to stop, look closely, and ask: what was really going on for the horse? When you read, watch, or listen to equestrians referencing historical practices—whether under the banner of classical equitation or positive tradition—practice discernment. Hone your ability to detect nuance, interpret the evidence, and choose wisely. Discernment is a thinking superpower—a blend of insight, ethics, and emotional awareness guided by observing critically the lived experience of the horse. Applying evidence-based knowledge—particularly The International society for Equitation Science First Principles of Horse Training—means, for example, that how we apply the aids must be grounded in Principle 9—Correct Use of Signals or Cues. Any consistent rider posture or body position that occurs simultaneously with primary aids can become part of the aiding system through a form of learning called compound conditioning (a subset of classical conditioning—see *Modern Horse Training: Equitation Science Principles & Practices* by Dr Andrew (McLean, 2024).

Understanding Principle 9 (ISES) (Principle 6 in Federation Equestre Internationale [FEI])—Correct Use of Signals or Cues—is crucial. It helps explain why some horses are labelled as 'dull,' 'lazy,' 'difficult,' 'hot,' or 'naughty.' These are not behavioural flaws; they can be as a result of unclear or conflicting signals from the rider. Coaching riders to maintain a consistent riding position becomes an act of discernment. Research shows that riders with better pelvic mobility and control produce significantly fewer conflict behaviours in the horse. In *Welfare Horse Sports*, how we apply knowledge is as important as what we know—procedural knowledge. Discernment is what allows us to move from knowing to doing better—for the horse, and with the horse.

1. *From Control to Collaboration*: Embrace evidence-based knowledge of how horses learn. Apply learning theory to communicate with clarity, expanding a horse's competency and resilience. Rather than controlling horses with tight nosebands, strong bits, and whipping, apply evidence-based knowledge. Shift from controlling horses to skills that provide horses with predictability and controllability of their world, not fear, pain, and confusion.
2. *From Static Expertise to Lifelong Learning*: Through the practice of curiosity, the skill of unlearning, learning updated evidence-based knowledge, and the application of equitation science within leave horses better (LHB) communities of practice (CoP).
3. *From Reactive to Proactive*: Earn public trust by safeguarding horses. Gain insights into the small deposits that leave horses better.
4. *From Fear of Forced Change to Trusting in Co-Creation*: Recognize evidence-based equitation (EBE) enhances the lived experience of horses, empowering equestrians to enhance horse welfare outcomes.

Practising shifts in our beliefs and attitudes starts with curiosity, followed by unlearning, learning, and finally the practice of reflective thinking before the cycle starts again, constantly repeating as we update our prior knowledge. In an era of misinformation and changing societal attitudes, this book is the equestrian's blueprint for positive change for horses. As Chapter 1 will detail, the equestrian world is experiencing a significant shift due to increasing societal scrutiny and a deeper understanding of horse welfare. This growing pressure necessitates a fundamental change in how we approach horse sports. Chapter 1 lays the groundwork for this transformation by exploring the urgent need for a welfare-centred approach, driven by evolving societal values and ethical considerations. It also begins to examine the ingrained mindsets and historical beliefs that often hinder progress.

Chapter 2 illustrates why a new perspective is essential for the future of equestrianism, paving the way for the welfare horse sports mindset.

Chapter 3 explains how the application of the Five Domains (2020) and First Principles (ISES) frameworks provides the evidence of welfare and ethical literacy, introducing how we leave horses better.

Chapter 4 explains CoPs, the impact of a LHB CoP with a practical strategy for communities to lower resistance to change and update prior knowledge with the application of Five Domains (2020) and First Principles (ISES).

1

THE SHIFT

Why *Welfare Horse Sports* now? Public trust is eroding, and equestrians face growing scrutiny from both within and beyond. Welfare breaches—once hidden in plain sight—are now broadcast and debated on the global stage. The public is watching, and increasingly, not liking what we see. Yet within this discomfort lies the possibility of reform. If we reimagine horse sport through the lens of equitation science, animal behaviour science, animal welfare science, and systems thinking science, we create a future not built on tradition, but on trust—earned through consistent, visible welfare improvements. What could horse sport look like if we centred the horse's lived experience? And what becomes possible when we do? You're reading this book, so I don't need to convince you of the case for welfare horse sports. Fortunately, it is not up for debate. However, as a curious equestrian, why is my book called *Welfare Horse Sports*? Without a collective witnessing of blue tongues, tight nosebands, hyper-flexed necks, whipping, and equine pain faces in different horse sports, I would not be writing about doing positive change for horses. This book is my invitation for us to explore actionable, observable, meaningful for horses, ways to leave every horse we interact with better. It is a blueprint for doing positive horse welfare and enhancing welfare outcomes for all horses.

Our obsession with the type of connection we have with horses (Jones McVey, 2023) coupled with a commitment to *do* better, because we now know better, has challenged equestrian attitudes towards the use of horses in sport (EEWC, 2022). Instead

DOI: 10.1201/9781003608066-1

of thinking about doing horse sports *with* welfare, this book is my invitation to leave every horse we interact with, better.

In Chapter 2, I present a welfare horse sports *mindset,* taking an intersectional approach (see Figure 1.1) to *how* we know what we know, *what* we know that optimises horse welfare outcomes, and *why* different ways we show up can leave horses better (mentally, not just physically). At the heart of how, what, and why is a *welfare horse sports mindset.* Updating the equestrian mindset is actioned by the practice of *reflective praxis.* Observable behaviours of a welfare horse sports mindset are evidenced by *doing* reflexive praxis (actions focused on introspection). Building upon the identified need for change in Chapter 1, this section introduces the core concept of this book: the welfare horse sports mindset. This mindset represents a fundamental shift in how equestrians understand and interact with horses in sport, moving beyond traditional practices

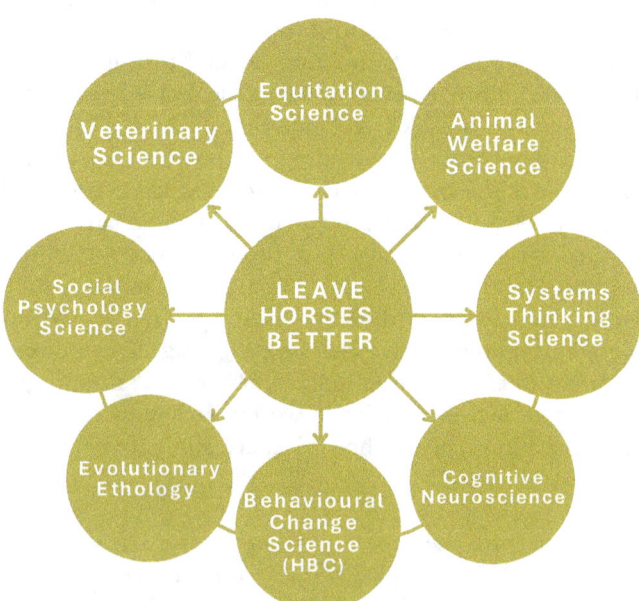

Figure 1.1 Whole Systems Thinking: The Heartland of a Welfare Horse Sports Mindset

to prioritise the horse's well-being in every aspect. It is the essential framework for responding to the 'shift' discussed in Chapter 1.

Attitudes are valence estimations of positive or negative qualities. Often, attitudes are mixed up with beliefs. Beliefs are estimations of whether something is or is not true. Attitudes are mental and emotional positions towards specific ideas, people, and situations. In biology, when organisms have the capacity to change but there is little encouragement to do so, the organism remains mostly the same from one generation to the next. But when the pressure to adapt increases, the pace of evolution increases in response. Over long timescales, patterns emerge: stretches of sameness punctuated by periods of rapid change. In horse sports today, we see this pattern, known as the punctuated equilibrium. This book is to help equestrians implement ethical horse training, using a welfare horse sports mindset to navigate through equestrian punctuated equilibrium—a rapid pace of change due to changed societal attitudes. When the human brain receives new information and feels uncertain, we use our current understanding, pieced together from previous experiences, to make sense of new knowledge. Brains do this via two mechanisms Piaget, J. (1952) The origins of intelligence in children. New York: International Universities Press. We try to assimilate new knowledge into our existing models. When assimilation fails, we feel viscerally uncomfortable and resist accommodation by trying to apply our current models of reality to the situation. It's only when we accept the existing models will not resolve the incongruences that it updates the model itself by accommodating the novelty. The result is an epiphany, the conscious realisation that our minds have changed. Great conversations encourage accommodation as everyone trades unique ideas and perspectives. Our mental and emotional position towards the use of horses in sport can no longer incorporate incongruences, and when we update our priors, it isn't the evidence that changes, but our attitude (mental position towards whether something is true or not), creating a paradigm shift. As our collection of attitudes builds up over time, they require us to seek different explanations for what we previously thought was settled and understood. At a certain point, what we thought was 'a good life' for animals is now a 'point of balance.' Our minds have changed. Sport creates a collection of attitudes, and this book provides the equestrian with up-to-date attitudes towards the use of horses in sport, enriching sport horses by *doing* on purpose, enhanced welfare outcomes, and leaving horses better, because of our interactions. A welfare horse sports mindset is a mental framework enhancing horse welfare outcomes. Like the growth mindset, a tool for learning and overcoming challenges, a welfare horse sports mindset influences your behaviour, decisions, and overall approach to horse welfare. By applying a welfare horse sports mindset, equestrians can individually make a difference towards horses living 'a good life.'

The welfare horse sports mindset is a mental model prioritising *learning*. However, it is the ability to update our prior knowledge (we did the best we could with what we knew at the time) through *practice*; procedural learning of evidence-based knowledge. The welfare horse sports mindset 'calls in' evidence-based knowledge *at the right time*, doing what makes an actual difference to the individual horse. Procedural knowledge requires lots of practice, having not learned the procedural part of knowledge. For example, identifying conflict behaviour (evidence-based knowledge) *in* competition (application of knowledge) is having a welfare horse sports mindset. Conflict behaviour was recently redefined by scientists as knowledge of learning theory (learning processes), biomechanics, and an understanding of horses' possible reactions to pain. By applying a welfare horse sports mindset, the multidisciplinary approach investigates the function of conflict behaviour ('calling in' pain) as stress-induced behavioural changes arising from conflicting motivations, especially when avoidance behaviour is prevented. In equitation, conflict behaviour may be caused by the application of simultaneous opposing signals such as stop and go cues, with the horse unable to avoid enduring pain/discomfort from relentless rein tension and leg pressures. The mindset of doing procedural learning to enhance horse welfare is *applying* evidence-based knowledge of conflict behaviour (bolting, shying, freezing, bucking, rearing, head-tossing, tensing up, tail swish, open mouth, tongue out, and teeth grinding), intersecting the function of the unwanted behaviour. The judge applying a welfare horse sports mindset recognises conflict behaviours *and* recognises that the behaviour is serving a function—an attempt *by the horse* to *resolve* their negative experience. Researchers studying the effect of the rider on conflict behaviour, rein tension, physiological measures, and rideability scores found riders significantly affect rein tension, heart rate, and increases in saliva cortisol concentrations, as well as specific conflict behaviours. This contribution to knowledge of the lived experience of sport horses highlights the impact *riders* have in enhancing welfare outcomes, with some riders inducing *more* discomfort in horses. A welfare horse sports mindset is a framework for approaching the use of horses in sport. Prioritising the skill of applying evidence-based knowledge at competitions, practising skills development, curiosity, unlearning, and procedural learning is the *welfare horse sports* driven mindset. It is a mindset guiding equestrian choices to enhance horse welfare outcomes.

Curiosity (willingness to challenge our assumptions and biases) and unlearning (traditional ways of thinking) to *renovate* (keep what works, replace what doesn't) horse sports is the right thing right now. At this moment in the arc of horses in sport, there cannot be a more important book for equestrians. A welfare horse sports mindset is a mental model of service and learning. Serving horses by asking *ourselves* 'how is my horse *feeling*?' Prioritising understanding horses is an attempt to satisfy *their* needs, not our own. Having a serving mental model (like abundance or problem-solving) is essential where collaboration and support are foundational.

Doing the right thing for the horse necessitates evidence of equine enrichment. Not just because horse sports are heavily commodified, or due to the fear that 'public fasting' equestrian sport will impact revenue. Not even the 'dancing horses with blue tongues' streaming into living rooms is *the* only reason to enrich horses. It is changed societal attitudes towards sentience and supremacy. There is a shift from 'being right' to 'getting it right' (for horses). This book is your roadmap to applying evidence-based integrative approaches to improve horse welfare outcomes. Instead of criticising different 'knowings,' we get to dive deeper, beyond tradition. The welfare horse sports mindset is founded on curiosity, exploring research in *diverse* domains. How we know what we know, a domain called epistemology, is the study of the origin and validity of knowledge. A belief is a subjective attitude towards a proposition that is true. Knowledge passed down/along from coach to coach, for example, 'use your inside leg to outside rein,' is a belief (subjective attitude) that inside rider leg pressure applied *with* outside rein pressure is going to improve your horse's 'way of going.' When riders use one leg pressure, inside leg, as a cue for the horse to accelerate (go response), it is likely the horse will fall out with his shoulders. Thus, bringing the shoulders back to the direction of travel by applying outside (light) rein pressure to the direction of the horse's midline corrects the falling out. Inside leg then outside rein, sequentially, aligns with the knowledge of how horses learn, specifically the mode of learning called *operant conditioning* (light pressure to cue a behaviour followed by the release of light pressure for the behaviour [or any slight behavioural change towards the desired behaviour]), demystifying 'ride inside leg to outside rein.' A rider's belief (a subjective attitude that the proposition is true) does not require introspection. Evidence-based knowledge demands introspection. Procedural learning, that is, applying evidence-based knowledge to horse training, aligns with the horse's cognitive and biomechanical realities. Reflexive riders (skilled in introspection) can challenge tradition and equestrian beliefs (knowledge). A welfare horse sports mindset is formed first from cultivating curiosity, the antidote to uncertainty, followed by practising the skill of unlearning, learning evidence-based knowledge anchored in introspection, and the skill development of reflexive praxis.

My experience of translating evidence-based knowledge (coaching equitation science) involves the importance of a 'grab' to motivate (motivational grab) riders to apply introspection. Introspection is anchored in reflexive praxis—where knowledge informs practice, and practice, in turn, refines knowledge. How we know what we know (epistemology) and understanding equestrian beliefs (subjective attitude towards knowledge is true) requires reflexive praxis competence; the skills and ability to identify reflexive practice skill gaps. Education in *procedural knowledge* recognises that 'inside leg to outside rein' is just one part of the required physical adaptations from the horse, on their unique performance journey that, may require the ability to collect.

The problem with holding a belief (a subjective attitude towards a proposition as true), is it does not require active introspection. Riders do not consider if the horse can even respond to both inside leg and outside rein pressure together. They cannot. However, it is the subjective attitude that both leg and rein pressure together must be true for the horse to improve, ignoring procedural knowledge that horses, like many animals, cannot respond to two pressures (cues) at once. An animal training phenomenon known as 'overshadowing' is when two cues/pressures are applied simultaneously, only the most salient (loudest pressure) will motivate the horse to respond. Overlapping leg and rein pressure fails the First Horse Training Principle (International Society for Equitation Science [ISES]), Principle 9: *Correct Use of Signals or Cues*. Unclear, ambiguous, or simultaneous signals lead to a mental state of confusion. This mental confusion can be seen as an observable behaviour to equestrians and is referred to by equitation scientists as conflict theory. A functional behaviour for the horse is attempting to resolve a negative experience. A conflict behaviour is a response by the horse to stress (in this example, confusion, but the cause of conflict behaviour also includes fear), helping the horse overcome/resolve the acute stressor. A welfare horse sports mindset identifies the observable response by the horse to the clashing aids and understands that the horse is simply trying to resolve his/her their confusion, thus procedural knowledge matters to enhance horse welfare outcomes, correctly identifying the observable behaviour as conflict behaviours, as opposed to projecting labels like 'naughty horse,' 'moody mare,' or 'a psychopath.' Rider/handler aids need to be easy for the horse to discriminate different responses, with each aid having only one meaning and all aids having different responses that are never applied concurrently. The final application of evidence-based knowledge (procedural knowledge) is an aid applied in time with limb biomechanics. For example, by applying an aid at the start of the swing phase of the limb, your aid is motivating a response. If the opposite happens, the stance phase, we just wait until the limb *can* respond to the rider's light aid biomechanically.

There were also beliefs held at that time in the world, where introspection was not required from equestrians. Few, historically, have carefully considered the horse's aversion to them, assuming the horse is the problem, framing their narrative that the horse's avoidance is because the horse is 'naughty,' 'moody,' 'disrespectful,' 'stubborn,' or 'lazy,' of which the horse is none. A welfare horse sports mindset identifies the negative experience *for the horse*, and understands the conflict behaviour (bucking, rearing, spinning, teethgrinding, tail swishing, napping) is the horse's attempt to resolve the negative experience. Taking an equi-centric perspective, I discuss later the telos (e.g 'horseness of a horse') as the core tennent of a welfare horse sports mindset.

The German term 'Wissenschaft' refers to the interdisciplinary nature of scholarly activities, recognising that knowledge spans multiple fields beyond what is traditionally considered 'science' in English. It bridges the natural sciences, social sciences, humanities, and formal sciences under one over-arching umbrella, the pursuit of knowledge. The heart of welfare horse sports; evidence-based knowledge, multidisciplinary approaches, and agility in addressing intersectionality in order to get closer to the truth about horses, leaving horses better than when we met them (positive change).

Horses were legally recognised as sentient in the United Kingdom as recently as 2022. Enriching sport horse lives is not just morally or ethically right. In Switzerland, France, Spain, Germany, New Zealand, and the United Kingdom, it is also legally right. The change in societal attitudes towards animals experiencing emotions as sentient beings ended a tradition where animals entertained humans for decades. While the circus in the United Kingdom died, the was zoo renovated, and today these thrive in the face of changing societal attitudes. Using animals *for* entertainment shifted theatrical performances to evidence of enhancing animal welfare and education. Captive animal collections (zoos) have been redesigning animal enclosures to align with species-specific behaviours (also referred to throughout this book as *telos*) for some years. Teleology (telos) is the needs of a species being satisfied, demonstrating captive animals can progress towards 'living a good life' by satisfying the needs of the species in captivity. Coined 'meaningful movement' enclosures, redesigning a tiger enclosure was the 'vehicle' that provided captive tigers with positive opportunities to thrive; navigation, novelty, agency to exert control, experience diversity, and behave socially (*www.carefortherare.com/meaningful-movement*). These 'meaningful movement' enclosures led to positive change in the lived experience of the captive tiger. Shaped by changed societal attitudes, satisfying the telos of a species is referred to throughout the book as having a mindset 'update.' Future-proofing (sustaining) the use of animals in entertainment is vital. Sea World updated its theatrical performances with welfare and education ripples; increasing visitor engagement and understanding of orca behaviour. Choosing to bring the functional complexity of a species' wild environment into managed environments was a mindset shift. In *response to* shifted societal attitudes, it is crucial to integrate the animals' needs *with* the needs of humans. Changing societal attitudes challenge the status quo. A triple bottom line strategy: planet, people, profit, is an example of the business mindset shift required to survive today in business to achieve long-term sustainability of an industry. A shift from entertaining to elevating animals' quality of life is identical to the welfare horse sports mindset facing equestrians. Ripples are created by doing positive change. In the ripples evolution research, positive change practices may never fully be measured. That said, ripples do result in *cascades* of change. Zoos have embraced societal attitudinal change with responsibility (ability to respond) and accountability for both the *mental* and physical states of animals held captive by humans. Zoos have worked collectively for the past decade, progressing captive animals towards 'a good life.' A 'good life,' as defined by the Good Life for Equids White Paper (2024), is a life balanced between *fluctuations* of positive and negative experiences through the life cycle of a horse, resulting mostly in positive experiences.

In a relatively short space of time, the digital age has positively disrupted traditional ways of 'doing life'; from blockbuster to Netflix, taxis to Uber, hotels to Airbnb, society has unlearned traditional ways of thinking in less time than dignity was granted legally to all relationships (legalising gay marriage). The post-pandemic societal and economic shifts, escalating climate change and impacts and the rise of artificial intelligence (AI) make today's pace of change unprecedented. As the digital era migrates into the infodemic era, the overwhelm and paralysis as information floods in (both accurate and false) creates confusion and moral uncertainty to *do* right by our horse. It is this democratisation of information, where anyone with internet access can access knowledge, empowers individuals, and at the same time proliferates falsehoods. The ease of spreading misinformation and disinformation destroys trust in organisations and media. Echo chambers and filter bubbles multiply online, reinforcing long-held and outdated equestrian beliefs that 'this is how we have always done it'—a certain companion in these changing societal attitudes. Captured eloquently by Major Timmis in his horse training book published in the 1930s; 'The average horse-dealer thinks no one knows as much as he does…to convince them they are in the wrong is impossible.' The welfare horse sports mindset, at its core, is the ability to think *critically*.

With the rise in quantity and quality of equitation science research, interdisciplinary research is making small and meaningful steps towards knowing what it is like for a (telos). A welfare horse sports mindset keeps thinking critically at the centre of this mental model (see Figure 1.1). Critically reflecting on peer-reviewed knowledge, the powerful emotion, regret, 'rears its necessary head' albeit at different times on our individual equestrian journeys. According to regret researchers (Pink, 2022), 'looking back' in order to move forwards is the power of regret. By enlisting our regrets, they can serve as motivation to change an immediate situation for the better. Pink urges ditching the cliche 'no regrets' and instead using the power of regret to *rethink* what we thought we knew and explore our ability to unlearn.

With pressure escalating (inside and outside of equestrianism) to address horse welfare gaps, why is an ability to unlearn and renovate horse sports so hard? It seems the struggle to rethink what we know is not a modern horse training challenge. The way cultural patterns are cued and encoded, like a domino effect, means horse advocates can learn how to initiate changes that harness positive ripples. Historical equestrian encoding to 'be the knower' made us act in ways that mesh the knowledge of the equestrian minds around us (groupthink). Humans have even evolved specialised brain systems to share knowledge within groups. If someone in your foraging band dislodged a coconut from the tree, learning by watching, soon the whole group shared this skill, resulting in clear coordination with each other, all following the shared script. Different groups around the world developed different pools of common knowledge, resulting in *different* cultures. Group membership to a culture makes a behaviour of the culture a behavioural norm, making the behaviour even more advantageous. As even more similarity reinforces the cultural norms; predictability and culture begins to experience a sense of 'us'—an expansion of identity beyond kinship to the broader group. If equestrians are historically coded (culturally reinforced not to change our mind or challenge what we know in the face of sharper logic or stronger evidence), it is easy to find empathy for reverting to our cultural coding *'this is how we've always done it,'* despite seeing horse welfare gaps exist. Without renovating for enhanced welfare outcomes for horses, there is uncertainty towards the future of using horses in sport, especially in light of immense societal change. However, welfare horse sports is a mental model bringing in the field of cultural evolution science. The field of cultural science explains culture does change across generations. The process of cultural evolution has been expressed in complex mathematical models; the central insight is that cultural transmission hinges on *learning processes*. As culture evolves, some elements are learned (patting horses) and others are overlooked (tightening nosebands) as the rising generation *recreates* society. Studies implicate biases in social learning, specifically conformity bias. Widespread behaviours are more likely to be learned (patting horses) than rare behaviours such as equestrians scratching (not patting) at the base of the wither, mirroring the behaviour of horse behaviour (mutual grooming). Cultural science explains that bias towards continuity with deep tradition inhibits cultural change. In addition, our impulse to mesh social approval/validation and peer instinct with backward-gazing nostalgia is part of our ancestral instinct. Finding comfort in tradition and the duty we feel to maintain traditions is a human instinct. These instincts are inside all of us; the conformist who seeks belonging and the traditionalist who cherishes continuity. According to research, a third instinct, the hero instinct, drives contribution and aspirations for esteem and success. Each instinct guides people in adaptive directions. Our shared equestrian culture (behaviours) for enhancing horse welfare brings together a myriad of individuals, ideas, and experiments for individual and community (pony club, riding schools, sport disciplines) *unlearning*.

In July 2024, the Fédération Équestre Internationale (FEI) unlearnt the ancestor instinct; 'this is how we have always done it,' publishing a Welfare Strategy born from the 37 Equine Ethics and Wellbeing Commission (EEWC) recommendations and accepted in full by the FEI. Just weeks later, tongues of equine Olympians were visibly blue, professionally captured by photographers in attendance at the Paris Olympic Games, and later published online. Vets and equitation scientists declared, 'organs turning blue is never a good thing' (Uldahl, as quoted in *The Times*, 2024). Holding up the ancestor instinct 'we have always done it this way' are the riders and trainers benefiting from the status quo. The peer instinct is uptaken in the absence of research (evidence), as opposed to erring on the side of the horse (a mindset that the horse is a sentient being). The absence of evidence is not evidence of absence, reminded to us by the EEWC following FEI riders, trainers, and some scientists persistently responding to the images of blue tongues with the statement 'we need more research.' This is in contrast to using the precautionary principle, where practices questionable towards the welfare of the horse are sought to be solved or discontinued immediately. Any incentivisation of pain (the winning of medals with blue tongues) or confusion (evidenced by conflict behaviours; mouth opening, tail swishing, spooking, and bolting) reveals incongruence between the use of horses in sport and FEI storytelling for many years, 'a happy equine athlete.' Unlearning the ancestry instinct; 'this is how we have always done it' is possible. The FEI launched the Be a Guardian campaign in 2024, as its 'vehicle to get the community on board' with its Welfare Strategy and Action Plan: 'In order to bring the community of riders, coaches and judges along with us,' said FEI President Ingmar de Vos.

This book is a blueprint for unlearning equestrian ancestry: the mindset of 'this is how we've always done it.' It serves as a companion for exploring how beliefs, attitudes, and mental frameworks shape our perception of horse behaviour and our response to the accelerating pace of change in equine welfare science. *Welfare Horse Sports* is presented here as a lens: one that sharpens our ability to be responsive to the horse and reflexive about our own practices throughout the horse's entire life cycle. At the end of each section, you'll find reflective activities titled 'Reflexive praxis.' The more deeply you engage with these, the more meaningful your contribution to improving welfare outcomes for sport horses, both individually and collectively. This mindset—a combination of responsiveness to the horse and reflexivity in ourselves—arises from a growing consensus: what truly matters is the horse's own perception of their physical and emotional state, from birth to death. A horse living a good life requires more than the avoidance of suffering; it requires the opportunity for positive experiences. That's the essence of welfare—or more fully, fairing well.

Welfare Horse Sports puts the horse's perspective first, (telos) honouring their mental and physical state. To the author's knowledge, this is the first resource of its kind to support equestrians in unlearning inherited practices and choosing instead to update their knowledge using evidence-based approaches. It's about replacing tradition with intention, and habit with humane science. For example, rather than expecting a foal to 'know' how to lift a leg on command, a *Welfare Horse Sports* practitioner understands how horses learn. They apply light pressure to motivate the response, release immediately once the leg lifts, pair this with a consistent auditory cue (like a cluck), and reinforce with a scratch at the withers—a species-specific affiliative behaviour (allogrooming). This layered, thoughtful interaction makes the horse more likely to offer the behaviour in the future, and with a positive emotional state. Instead of simply 'adding welfare' to sport—like applying a noseband taper gauge to prevent excessive tightness—*Welfare Horse Sports* asks us to reimagine sport itself. It seeks to design positive experiences from the start, not just mitigate negative ones. And if you're still asking 'So what?'—the answer is simple: because it's the right thing to do. Because welfare science tells us we must. And because when we make decisions through the lens of what it's like to be a horse—not a human projecting onto a horse—we naturally begin to rebuild public trust. Trust we will need if we are to have a future for horses in sport at all.

Welfare is:

> The state of the animal's body and mind, and extent to which their nature is satisfied.
>
> Fraser et al. (1997)

Our horses cannot experience having their species-specific needs satisfied if we still believe that welfare is care. Before engaging in a welfare conversation, ask what is *their* understanding of welfare? Actively listen. Not to respond. If you hear what you think sounds like care—a basic minimum physical state (may include mental state, i.e., negative emotions)—then discussing opportunities for horses to experience positive mental states; joy, curiosity, etc. may be a few more conversations away for the time being. Later, I share how we can try and meet horses and humans where they are, instead of educating at people. However, if you do find yourself in any discussions where replacing welfare with 'well-being' is gaining momentum (due to negative associations with welfare), ask this one question— To what extent is the horse's nature being satisfied? Welfare is the physical state and performance is important, so too is the horse's subjective experience—as perceived by the horse. When we recognise that 'welfare' means different things to different people, just knowing this can, in some instances, instigate positive change for horses.

 REFLEXIVE PRAXIS

The point of learning is not to affirm what we currently believe. It's to evolve what we believe." (Grant, 2025). Steer discussions from removing/avoiding the word *welfare* in our community spaces, in person and online, to inconvenient conversations that help identify that welfare is beyond the physical state of the horse. Ask to what extent is a horse's nature being satisfied? The telos of the horse (horseness of a horse). This calibre of conversation sprinkles 'seeds' of trust, evidenced by the different objective ways (evidence-based knowledge) the needs of horses are being satisfied.

Welfare is a physical state and performance is important, but so too is the horse's subjective experience—as perceived by the horse. The field of animal welfare science has agreed that welfare is: Today, society places its spotlight, rightly, on both the physical *and* mental condition of animals. We have expanded our consideration 'bandwidth' to the experiences (as perceived *by the horse*) *equally* alongside physical state, together, offering *their* welfare state. Welfare, the word, is necessary because it represents the individual's *mental* state. The extent to which a horse's nature is satisfied requires us to update our priors about species-specific knowledge; ethology, cognition, anatomy, and physiology. Learning (unlearning) the updated (evidence-based) knowledge of the horse takes equestrians closer to understanding how the horse views the world. A welfare horse sports mindset is making decisions towards enhancing the mental and physical state of the horse (welfare) by providing opportunities for the sport horse to experience *telos*, the Greek word for fulfilment. A welfare horse sports mindset is intentional choices towards satisfying the specific needs *of the horse*. Why does having a fulfilled life matter? Or perhaps the better question is—'So what?' Considering the cumulative life experiences of the sport horse is necessary when assessing the quality of life of the sport horse.

Originally developed to assess the welfare of experimental animals, assessing quality of life is relevant to all animals today, with some arguing even more relevant to the use of horses in sport. Quality of life is becoming increasingly important in animal welfare and is not a one-off welfare assessment. The Quality of Life Framework provides an insight into the balance of all the negative and positive welfare experiences a horse might have during their lifetime. Three categories for quality of life have been recognised (Farm Animal Welfare Council [FAWC] report): a life not worth living, where the negative experiences outweigh the positive—the animal would be better off dead than alive; a life worth living, when, although some pain, suffering, distress, or lasting harm may occur, on balance, the animal's positive experiences outweigh the negative; and a good life where positive experiences heavily outweigh the negative beyond any reasonable doubt; certain husbandry practices are proscribed and others prescribed.

Quality of life, determined by the balance between positive and negative experiences over time, means focusing on helping resolve negative experiences is of value, as providing opportunities for positive experiences.

> 'A Good Life' is the balance of salient positive and negative experiences is strongly positive. Achieved by full compliance with best practices advice, well above the minimum requirements of welfare.
>
> Eurogroup for Animals (2024)

A good life for the sport horse, I believe, is possible when we show up to identify accurately (without euphemisms) the behavioural expressions of the horse, what our horses are communicating, when we help get their needs met, *and* when their experiences of interacting with us are predominantly positive. When thinking about a good life for a sport horse, it is helpful to visualise fluctuations or 'shifts' between three emotional groups in which all emotions can find a group; negative, neutral, and positive. Throughout a 24-hour period, the ability of the sport horse to move through negative experiences, for example, compete for 5 minutes in an arena, is essential. One negative experience for the sport horse facing this scenario is the absence of their 'best friend.' Either their BFF is back at the stables, maybe on-site, or back 'at home.' Learning to see the competition experience through the lens of the horse, that is, develop your welfare horse sports mindset, is developed by knowing the species-specific needs of horses, including how their sensory inputs differ from ours. Yet it has been common practice for sport horses to perform without having their species needs (proximity to other horses, sounds at showgrounds) resolved. Applying the welfare horse sports mindset means making decisions to *resolve* negative experiences for the horse, not the human. A horse sport *with a* welfare mindset is when we notice the sensory responses of the horse (conflicted behavioural expressions) to the sounds at a showground (PA system, etc.) and resolve them as humans by adding a sound-muffling hood. Originally designed to keep flies away, competition riders have self-selected the ear bonnet or fly hood as a compulsory item of competition horse clothing. Manufacturers boast that 'dense soundproof micro-foam to absorb sharp or loud noises help with increased concentration and relaxation.' Knowing the lowest frequency detectable to horses is 50 Hz, which is higher than the lowest human detection threshold of 20 Hz, whilst horse hearing exceeds the highest frequencies heard by humans (33 kHz compared to 20 kHz for humans), is the signpost that doing welfare horse sports requires resolving/responding to the level and number of sounds simply unheard by the human ear. Even the funnel shape of the equine ear provides an acoustic pressure gain of 10–20 dB, elevating the acuity of horse hearing. Having a welfare horse sports mindset not only identifies sensory differences between both species (horse–human) but uses evidence-based knowledge to resolve for the horse aversive sounds, resolving for humans dangerous, unwanted behaviours to the sounds at showgrounds. It is a mindset of responsibility for the lived experience of the horse, a different species from humans. Throughout this book, you are invited to shift or expand your welfare horse sports mindset and take on the responsibility of providing a good life for sport horses, through your responsivity to the horse's actual lived experience (via curious objectivity) and reflection on your responsibility in helping horses to live a good life (confident humility).

In para-dressage, horses experience competitions with a 'buddy' horse. The close proximity of the 'buddy' horse may offer the opportunity to experience positive emotions (affect) for these sport horses. Going beyond visual proximity, the ability for physical touch has the potential to provide positive effects. The final action in this example of a welfare horse sport mindset is to provide post-performance the opportunity for the sport horse to have a positive experience. The sport horse is offered the opportunity to touch and allogroom, if desired, post-performance. The behavioural signs indicative of pleasant or unpleasant experiences have been recently systematically reviewed, offering our welfare horse sports mindset a checklist of behavioural expressions to aid interpretation of the experience for the sport horse. Responding to the behavioural expressions of the horse to human control and training, as negative or positive affective (emotional) states, is an opportunity to enhance horse welfare.

Fluxing throughout the day, moving through a negative experience (confinement) into resolution freedom to roll in a sand bath (positive experience) into rest (neutral) is how our mindset helps demonstrate welfare horse sports. Like the rays of the sun between the hours of 12 pm and 2 pm, which are harmful for humans (in some instances creating harmful mutations such as skin cancer), we mitigate for ultraviolet radiation (UV) by applying sunscreen and/or removing ourselves from the rays by staying in the shade for this duration. What we have not done as a species is remove the sun from the sky. Mitigating potential negative experiences, such as applying sunscreen, starts with our mindset. By identifying the causes of negative emotions in the sport horse—pain, fear, confusion—we can then set to resolving (the framework of the welfare horse sports mindset) these negative experiences, that is, tack (saddle and bridle) comfort, veterinary investigation, and the correct application of learning theory, specifically removal reinforcement for the desired behaviour and providing clarity of communication for horses through the application of the science of how horses learn. Removing horse confusion to the rider/handler's aids/signals/cues is crucial. When this does not happen, the horse feels confused or fearful, and we shape back down to the previous stage of learning until the horse is sharing behavioural expressions of confidence in the previous behaviour. Being able to shape up and down *in response* to your horse's behavioural expressions through your interactions can lead to avoiding the 'culmination factor.' All the small negative fluctuations throughout the day, if they are not resolved by the horse, culminate. This can look like avoidance by the horse of humans, and the opposite. If your horse is absent of friends, forage, and freedom (the three Fs), when we arrive, being the only interaction for the horse, the horse will approach you *because* of the absence of stimuli in the environment. A culmination of small negative experiences (absence of resolution throughout the day) shows up in the horse's mental (emotional) state. Finally, offer opportunities for positive experiences, such as replacing patting with scratches at the base of the wither after jumping a round of fences, on landing out on cross-country, or at the final halt on the centre line of the dressage test.

Actioning a welfare horse sports mindset is about doing small accumulative change for the enhancement of horse welfare outcomes. By making continual incremental improvements, we do positive change for the lived experience of the individual horse (welfare). At the core of our mindset is:

- Identify negative experience/s: *Reflection*
- Reduce or resolve negative experience/s—by dealing with obvious compromises or risks: *Resolution*
- Provide opportunities for positive experience/s—by facilitating as many positive experiences as possible, culminating in the horse feeling safe and secure (interactions with you are predictable and controllable), spending time in the company of friends (one where deep relationships are experienced, e.g., mutual grooming), feeling the vitality of health and fitness, and living with a sense of agency to choose what to do, when, and with whom: *Enhancement*

IS YOUR HORSE LIVING 'A GOOD LIFE'?

A GOOD LIFE? CHECKLIST

- Need for companionship
- Need for forage: 3 Fs prioritised from birth, throughout life, and into death
- Need for space

THE THREE Fs

FRIENDS

Reflection

- Reflect on your horse's social life. Does the horse have consistent companions? Are there signs of friendship (bonding) such as mutual grooming? Journal about observed interactions and consider which horse/s have negative behavioural expressions in the turnout group.

- Observe how your horse interacts with companions during turnout. Look for body language that indicates positive or negative experiences (e.g., relaxed grazing together vs. pinned ears and chasing).
- Reflect on past introductions of new horses to the herd. Were there unnecessary stresses or conflicts?

Resolution

- Create stable friendships by introducing gradually a 'social window' created in the stable more natural social (group) environment.
- If signs of social behavioural stress (pinned ears, avoidance), explore experimenting with different individuals joining the group.
- If a horse appears socially isolated or distressed, reorganise turnout groups or reach out to an equitation science consultant about herd dynamics.
- Adjust future introduction strategies based on lessons learned, such as gradual exposure via introduction of social window stabling, and/or through adjacent paddocks.
- Establish steps in the future to introduce new herd members or optimise existing dynamics.

Enhancement

- Facilitate group activities, such as spreading forage in areas where horses can share.
- Establish protocols for introducing new horses to maintain social harmony.
- Organise opportunities for horses to interact naturally, such as *shared* grooming stations.

FORAGE

Reflection

- Is the amount of roughage in the diet, on a dry matter (DM) basis in relation to bodyweight—the minimum is 15 g of DM per kilo bodyweight (BW) and in addition to supplementary feeding? Greater quantities and varieties can be offererd.

- Is diet balanced by an equine nutritionist?
- Trickle feeding evidence. Check the ration lasts until next feeding time. If using slow feeding systems, check whether horse can ingest sufficient quantity.
- If fed concentrates, are they presented without restricting forage intake requirements? The horse should have either free access to or a constant supply of forage?

Resolution

- For pastured horses, where grass is poor, in short supply, or very green (low dry matter), pasture should be supplemented with sufficient quantities of a higher DM forage such as 85% DM hay.
- Check grazing quality and management.
- Check length and type of fibres (longer than 2.5 cm recommended, except for horses with dental problems).
- Can see other horses and their surroundings while eating.
- Can avoid conflicts over food (herd dynamics).

Enhancement

- Forage feeding system allows all horses to obtain sufficient quantities at the same time, and encourages a natural feeding posture.
- Evidence that the horse finds the varieties of foods offered palatable and shows interest in feeding.

FREEDOM

Reflections

- Evaluate your horse's daily routine. How much time does the horse spend confined? Is movement restricted unnecessarily?
- Watch for signs of frustration, boredom, or stereotypies (e.g., weaving, cribbing).
- Consider seasonal changes in forage availability. Did the horse experience abrupt changes in diet that caused stress?
- Review past turnout patterns and note any periods of excessive confinement.

- Develop long-term goals for your horse's environment. Can you increase access to tracks, varied terrain (track system), or provide sand baths?
- Develop long-term goals for your horse's environment. Can you increase access to trails?

Resolutions

- Address the root cause (reduce stable time) and plan for extended turnout in areas with engaging terrain; track system, sand bath, etc.
- Identify and address management constraints that led to reduced turnout.
- Address barriers like fencing or rescheduling constraints that limit freedom.

Enhancement

- Create enriching spaces in the turnout area with varying surfaces, logs, hills, and sand to encourage exploratory behaviour.
- Develop a turnout schedule that prioritises extended periods of movement, even in challenging weather or circumstances.
- Resolution action: Address barriers like fencing or scheduling constraints that limit freedom.
- Enhancement: Create a space that encourages self-exploration and exercise, such as a track system or pasture with diverse features.

Reflexions

- End each week by journaling observations under these categories.
- Reflect on how well you are balancing the 3 Fs and plan adjustments.
- Seek feedback from others involved in your horse's welfare
- By integrating reflexive practice into the daily and long-term management of your horses' physical and mental state, you can effectively reduce negative experiences, resolve risks, and amplify positive welfare outcomes.

Being responsive to the horse's lived experience is welfare horse sports' 'north star.' Like a muscle, practise the skill of being responsive to how your horse feels in all your interactions. Unlearning 'what I have always done' can ignite curiosity to learn evidence-based knowledge. Similar to being notified by your phone that an update is ready for installing (fixing any bugs to run smoother and faster), a functionality mindset upgrade is also referred to by McLean (2024) as 'What is it like for a horse to be a horse?' a modern take on the philosopher Thomas Nagel's essay 'What is it like to be a bat?' The essay concluded that this is the wrong question. What we should be asking is what is it like for a bat to be a bat? Research into the beliefs, attitudes, and values of equestrian enthusiasts and practitioners attributed the inability to make changes to horse welfare not to themselves, but to external circumstances beyond their control. The human condition for certainty through the need to control has sparked interesting research to improve horse welfare. Professing our love for the horse, a study involving equestrians identified an absence of recognition of the conditions under which we keep and use horses, as falling short of participants' own standards. Equestrians attributed the inability to make changes not to themselves but to external circumstances beyond their control. These results pave the way for further research to determine whether equestrian activities are based on the lived experience of the horse or if belonging to the equestrian world takes precedence, even if it costs the horse negative welfare.

To ensure horse interaction with humans does not become an increasingly negative experience, a commitment to consistently apply learning theory and responding to positive and negative behavioural responses from the horse is at the heart of earning public trust. The use of equipment restricting the horse's behavioural expressions will also restrict the ability of equestrians to respond (response-ability) and is likely to substantially increase the aversiveness of the interaction. Renovating (keep what works, unlearn what is not horse-centric) is done by looking back at our long-held beliefs. Relentless pressures, high pressures, and simultaneous pressures should not have been winning prizes whilst simultaneously telling the public how 'much we love horses.'

Admitting that we don't know something is not a weakness; it shows integrity. Believing and pretending we have all the answers is derailing our ability to earn public trust. To get closer to the truth about horses and how they feel about horse training, at home and in competitions, start by thinking about choice, also referred to as agency.

 REFLEXIVE PRAXIS

- Does your horse choose you? Consider your horse's behaviour during key moments: when you approach to catch them, do they walk with confidence, curiosity, or show hesitation? When they see their tack in the stable, do they try to avoid you or remain calm and curious? The mounting block represents a pivotal moment—the first visual and physical cue in their experience of being ridden. Reflect critically on your horse's responses during these interactions. What does their behaviour suggest about their feelings toward you and the training you do together? Are they curious, neutral, or avoidant? Use these observations to assess your relationship honestly, and consider what changes might improve their experience and motivation to engage and be curious about you in their immediate environment.
- Would your horse choose your training methods? This is an intentionally challenging question designed to probe the core of your motivation. Why do you choose to train the way you do? If you believe your methods are hard but necessary, take a moment to reflect: why do you see them as necessary? Could there be another way?
- Imagine a 'third way'—a balance between what feels right to you and what might feel right to your horse. This reflection invites you to question assumptions, explore alternatives, and seek a deeper knowledge in your horse training content and the potential impact on the mental state of the horse.

Summarise your thoughts after this reflexive exercise—What did you discover? What might you change?

THE WELFARE HORSE SPORTS MINDSET

Think of unlearning as a critical *rethinking* of the 'ingredients' that shape our beliefs and attitudes. The belief that knowledge always accumulates doesn't apply when trying to ensure horses live a 'good life.' A welfare-centred approach requires letting go of outdated beliefs, such as a pat on the neck of the horse being inherently rewarding. Whilst a pat releases rein pressure, researchers found that repeated pats cause the horse discomfort, and replacing patting with scratches at the base of the wither mirrors mutual grooming, a species-specific behaviour that the horse finds rewarding.

Unlearning requires making room to question—even release—practices that do not align with enhancing welfare outcomes. When we cling to knowledge that hasn't stood the test of time (or science), we undermine equestrian credibility; we can be trusted to enhance horse welfare, even when no dressage judge is watching.

An example of unlearning is our adaptation to new technology. Consider adaptive cruise control in cars: when we first try it, we must unlearn habits like reaching for the steering wheel when it adjusts automatically. Similarly, we must unlearn outdated horse training techniques that don't adhere to the First Principles International Society for Equitation Science (ISES), even if we once believed they were beneficial.

In equestrianism, we often hear 'I've always done it this way.' But with society shifting, we're seeing more 'ethical' and 'welfare' language in horse sport communities. Instead of simply accumulating knowledge, equestrians should make room to critically reassess what we know. How do we know what we know? This question touches on epistemology—the study of knowledge itself.

Do coaches recognise the horse's role in training? According to Swedish researcher Zetterqvist Blokhuis (2021), coaches failed to respond to the 'voice' of the horse during a coaching session. JonesMcVey explains that ethical horse training requires attentiveness to the horse's affective state (emotions) and our responsiveness to the way the horse feels throughout the training session.

To truly be responsive, we must unlearn practices that fail to improve horses' lives. Reliance on relentless pressure, coercion, ill-fitting tack, and punishment (including tight nosebands) must be re-examined. Today we have growing scientific evidence of the importance of agency and opportunities for horses to experience positive emotions. We must challenge emerging stories labelled 'ethical' and 'welfare-focused' if claimed without the focus on responsiveness to the horse's inferred lived experience.

Unlearning doesn't mean abandoning foundational knowledge or discarding important skills. Think of it as an extension of learning: begin by asking which practices truly leave the horse better and which need updating. To facilitate a culture of unlearning, we can create spaces that invite self-reflection and questioning, free from incivility and judgement. Online communities such as Coffee with Horse Lovers offer equestrians a safe environment to explore unlearning. By embracing unlearning, we not only open ourselves to evidence-based training but also shift equestrian culture to one where the welfare of horses is prioritised.

In the end, unlearning is about returning to the horse's lived experience and making decisions that benefit the horse, not the human. Unlearning is about creating an environment where choices are guided by how the horse feels, not just what we can achieve. This shift challenges equestrians to look beyond 'How we've always done it' and ask instead, 'How can we do better?'

Through unlearning, we can take our next steps towards welfare-centred (equi-centric) equestrianism. In the words of Albert Einstein, 'the measure of intelligence is the ability to change.'

Of course, in order for us to come out, there has to be a closet. And in the case of curiosity, the closet is the desire for equestrians to KNOW things. Are we here to be right? Or to get it right for horses?

 REFLEXIVE PRAXIS

Going into the 'red zone' in order to win a 1.60 jumping class (Guerdat, 2025) was explained by Olympic gold medallist Steve Guerdat as jumping a seven-stride distance (upright to an over) on six strides. The seven strides maintain confidence; the six strides riders are choosing the 'red zone' (for the horse). When riders make decisions to win the class over an optimal mental state (build confidence), is it time the sport renovates course design? Should we keep what works and change what is not *for horses*? Is it time to rethink (place the horse at the centre) what success looks like in the sport of show jumping? Can we start to rethink jump classes? Should we provide dressage competitions where success looks like lightness, release of pressure (for correct responses), and self-carriage (self-maintained speed, line, and outline), most of the time? How can horse sports renovate so the lived experience of the horse is centred? One way is to demonstrate responsiveness to a horse's arousal, anxiety, confusion, pain, and discomfort—signalled by the horse, for example, when the horse opens their mouth to alleviate the pressure down the reins.

Rethinking horse sports with welfare is the heartland of giving horses the opportunity to have *positive experiences* (third central tenet of the Five Domains). Exploring welfare in horse sports lands fairly easily for the 'other 23 hours of the day.' But what about *during* our horse–human interactions? The ability for riders to provide horses with lightness (the horse is motivated to remove the tactile pressure of a fly; otherwise, horses would not be motivated to flinch the skin as a consequence of the fly landing), thereby removing the lightest of pressures (a fly). The ability of riders to give horses what they need to *learn involves* the *removal* of the rider's light pressure, for more of the behaviour we want (this is how learning takes place, mentally facilitating *the experience of competency* for the horse), and then regular testing of training (the opposite of holding horses with constant pressure) with self-carriage tests (keep going until rider signals otherwise). When a horse experiences a mismatch between what she expects to happen (removal of rider pressure for deceleration) and what actually happens (rider keeps the pressure on), the experience for the horse is known as a *negative* prediction error. To help understand prediction errors, a fundamental concept in neuroscience explains how our brains (horses and humans) adjust our behaviour based on experience. If an event unfolds as expected, this experience is described as a positive prediction error. In this field of research, the most frequent animals studied and experimented with are mice, and the measurement of prediction errors is the measurement of the neurotransmitters during the behaviour of mice. When the outcome is different than what is expected, the brain uses prediction errors to update future expectations and guide learning. As riders, when we anticipate a horse will respond in a certain way to our aid/cue, and the horse reacts differently, thus indicating a prediction error. What is essential to giving horses positive experiences in equitation is for the horse to experience the release of the neurotransmitter dopamine, because the rider provides *positive prediction errors* (as expected outcomes). Providing the release of a light aid provides an opportunity for a positive emotion (specifically the release of dopamine) because the rider delivered what was expected by the horse—the release of the rein/leg pressure. In the field of neuroscience, the data on animals receiving predicted outcomes provides researchers with evidence of positive emotions (presence of dopamine, a neurotransmitter responsible for feeling good). Reflect on your ability to release pressure for your horse, for the correct response (release because of the desired behaviour), providing your horse with a positive prediction error—a small spike in dopamine (feel-good brain drug). However, when there is a different outcome from the one the horse is expecting, your horse also adjusts their behaviour to update future expectations.

RELEASING PRESSURE PROVIDES POSITIVE EXPERIENCE

Understanding prediction errors (neuroscience of training) requires you to understand the impact prediction errors have on how your horse feels. That requires reflective thinking. Does your training provide the outcomes predicted by your horse? Lightness? Release for desired behaviour? Self-carriage? If so, you are delivering positive prediction errors and, with that, small spikes of dopamine. Let us frame this nugget of neuroscience knowledge as a switch. I am using this simple framing to help more riders have the confidence to experiment with prediction errors *during* training. Using the analogy of a light switch to frame understanding prediction errors, when riders/trainers do a behaviour expected by the horse (removal of light pressure is the clearest for others to see, though also delivering light aids and testing for self-carriage are learned as predicable behaviours of the rider), you are *doing* positive experiences (release of dopamine) for horses, the third central tenet of the Five Domains (see Figure 2.1).

 ## REFLEXIVE PRAXIS

Recall a situation where your expectation differed from the actual outcome. How did this experience influence your subsequent understanding or behaviour?

The Five Domains (2020) is an update from the 2015 version. Most notably, 'Interactions' is added after Behaviour (Fourth Domain)—4a: Behavioural interactions with the environment, 4b: Behavioural interactions with other animals, and 4c: Behavioural interactions with humans. The Five Domains is a welfare assessment tool that makes the complex feasible by working systematically and logically through the relevant aspects of the horse's lived experience.

Domain 1 includes the conditions and experiences that relate to the horse's nutrition and hydration.

Figure 2.1 The Five Domains by Cristina L. Wilkins (Reproduced with permission.)

Domain 2 includes those that relate directly to the physical environment the horse finds themself in.

Domain 3 covers the conditions and experiences that relate to the horse's health and fitness.

Domain 4a: The conditions and experiences relating to their interactions with the environment.

Domain 4b: Those that relate to their interactions with other animals.

Domain 4c: Their interactions with humans. This initial breakdown into four domains guides you through a systematic, step-by-step assessment of the horse's lived experience. It helps you think carefully through the process in a rational and logical way.

Applying the model then requires us to deduce the horse's perspective. We are to reflect on and determine whether the horse is in a negative state or a positive state. It is your evaluation of negative and positive states which is used to represent their welfare state. It is very important that you work on one domain at a time. Your aim is to work out the horse's experience and welfare state in relation to one single domain. In other words, your welfare assessment is not about awarding an overall welfare grading for the horse's entire lived experience. This is because what matters to the animal is that you identify what is going well and what isn't, and that you learn precisely which conditions and provisions are leading to the positive experiences and need to be maintained, or leading to negative experiences that the horse needs to resolve. This approach is much more meaningful and helpful to the horse as their guardians.

FIVE DOMAINS (2020) AND FIRST PRINCIPLES (ISES)

Applying the Five Domains of Animal Welfare is a path to the possibility of horses living 'a good life.' The World Organisation for Animal Health (WOAH) acknowledged the significance of the Five Domains (2020) for evaluating the welfare needs of animals under human control in their vision paper titled 'Animal Welfare: A Vital asset for a More Sustainable World,' published January 2024.

A good life is a balance of positive and negative experiences, landing for the horse, strongly positive. Building a reflexive practice enables us to *apply* the latest in evidence-based knowledge, delving into equestrian practices that *improve* horse welfare outcomes.

Demonstrating that a captive animal can progress towards 'living a good life' by meeting species-specific needs, or telos, is essential. Coined 'meaningful movement' enclosures, these redesigned spaces provide tigers a 'vehicle' to learn to navigate, experience diversity, novelty, and agency to exert control, and behave socially where appropriate (*www.carefortherare.com/meaningful-movement*). Meaningful movement enclosures lead to positive change shaped by changed societal attitudes towards the use of animals in entertainment. Sea World shifted from theatrical performances to welfare and education, reporting an increase in visitor engagement and understanding of orca behaviour.

The consensus amongst animal welfare scientists with nuanced perspectives in the field is that welfare is how an animal experiences *their* life, physically *and* mentally. Welfare is the overall mental state experienced, which we infer from objective welfare indicators (Littlewood et al., 2023). Some animals in the world from progressive countries have even been granted legal personhood, giving them some basic legal rights. Lubelska-Sazano (2024) reviewed animal welfare law in Europe, identifying the amended Article 2.1 of the Dutch Animal Protection Act, which states that 'animals will no longer be adapted to the system, but the system will be adapted to the needs of animals' (Dutch Animal Protection Act Amendment, 2021). The United Kingdom, often hailed as a world leader in animal welfare laws, dropped from an A grade (highest rating) to a B grade (second highest rating) on the World Animal Protection Index during this same period (2020), and at the same time as the latest iteration of the Five Domains was published. The Five Domains (2020) collates animal *and* situational-based indicators (such as the tightness of a noseband) and takes responsible horse sports (international, national federations, and governing bodies pledged to apply the Five Domains 2020 Framework) through three central tenets of the Five Domains:

- Identify negative experiences for the horse
- Resolve negative experiences for the horse
- Provide opportunities for positive experiences for the horse

> Justice, that brightest adornment of virtue by which a good person gains the title of good.
>
> Cicero (1913)

As an pracademic, one of the most common questions that have been asked over the years is: 'How do I motivate my friend to change?'

The reality is, people only change when they feel like changing. It doesn't matter how much you want your friend to change, thereby enhancing her horse's quality of life, or that you are right; her horse will become the horse she needs. Nor does it matter how big the consequences are for her horse if they don't change. Using the welfare horse sports mindset, you know you can't make a rider change their behaviour. You can't make someone else change. But you can influence them.

When we pressure our friends (or strangers on the internet), it just creates more tension, resentment, and distance in our relationships. It can even do the opposite; as we push our friends to make decisions to enhance horse welfare outcomes, we make our friends (or strangers) push back. My own experience has found friends doubling down on their original behaviour towards their horses. When our freedom to choose (agency) feels threatened, we all activate our resistance to the change, in some circumstances intensifying our commitment to original behaviour. In short, our passion to enhance horse welfare is the spark for more of the same behaviour we were originally seeking to change. The researchers in the field of neuroscience found that when we push someone, it only makes the person push back. People need to feel in control of their decisions. We want horses to have a life worth living, so when we pressure our friends to change how they are training their horses, with the best intentions of course, it is actually yielding the worst results for horses. Because every time we take on human nature to feel in control with our 'righting reflex,' we lose.

Using the welfare horse sports mindset, you'll learn a new approach to dealing with situations where you want an equestrian to change *their* behaviour and enhance horse welfare. Here is the rub. *We* (you and I) cannot make someone else change, but we can *influence* change. How equestrians approach enhancing horse welfare starts with us—self-leadership. Modelling the behaviour we wish to see in others, we must keep our relationships and the equestrians who may frustrate and challenge our hopes for better lives for horses in mind.

'I wish you would see your horse's *naughty* behaviour as a way of solving their negative experience.' After hinting, sharing data (papers), and offering coaching, maybe you have been motivated like me to take my frustration to socials. Over the years, I have experimented with all different vehicles to communicate science (how horses learn) to enhance the horse's lived experience, *because* of our interactions. I've done the passive-aggressive comments, signed the friend up to a webinar on her behalf, and provided free spots to spectators at my equitation science clinics. If any were going to influence change, they would have. So I looked inwards, at myself. I turned to introspection. A welfare horse sports mindset practises compassion. Everyone is doing the best they can, with what they know at this time. Wanting a good life for horses (Quality of Life framework) is fantastic, but not at the expense of increasing resistance to change amongst equestrians. For horses to flourish through the experience of predominantly positive mental states, we must expand resilience by providing horses with predictability, controllability, and self-carriage in equitation. Positive horse welfare goes beyond ensuring good physical health and the prevention and alleviation of suffering. A positive mental state is the sum of rewarding experiences, including having choices and opportunities to actively pursue goals and achieve desired outcomes, according to the horse's telos and individual capabilities. Wanting your friends to provide their horse with a positive horse welfare life is normal. The issue isn't wanting this for horses; the issue is how we have been approaching human behaviour change (more about this later) and how we act towards our friends, is impacting horse welfare. Maybe as you are reading this, you are realising that someone has been pressuring you to change? They don't even need to say something; it is very clear they disapprove of your horse's current lived experiences. They want a different way of living for your horse. And your natural inclination is to push back. This knee-jerk reaction to pressure equestrians, and their innate reaction to resist, is why it is hard for anyone to change.

Just pointing out what the horse needs to enhance his experience, we assume it would be easy for them to just do it. All you have to do is point out the obvious, right? I have definitely been there. Just tell the rider how much better the horse will feel about being restricted (stabled) when you give them access (via a social window) to their best friend (pair bond). I mean, clearly they've never considered this option for themselves, right? Now, flip that around. How many times has someone pointed out the obvious to you? As if you didn't know that giving your horse access to their best friend would change their mental state? Or that releasing pressure for the correct response (pressure down both reins) would solve for your horse the negative experience of bit pressure? It is almost offensive when someone else does this to you. You feel attacked. Worse, when the annoying person telling you what you must do elaborates the righteousness of their verbal tone, acting like it is easy to change after decades of habit formation; keeping hold of the reins, despite the fact that your horse is now immobile (in halt). The fact is change is hard for everyone, including you. No one wants to feel pressure. Of course, equestrians (horse lovers) want their horse to live a good life. Change is never easy. If it were, equestrians would already be making the change; horses would be flourishing because of their resilience and fortitude and non-equestrians (the public) would approve of the use of horses in sport because our sporting activities offer horses enrichment opportunities. The best thing all of us can do is to stop pressuring our friends to change. We all have completely unrealistic expectations of our friends to change, and our current approach is backfiring. Let us turn to the science of motivation and change so our approach to enhancing horse welfare is more effective. Stop trying to motivate equestrians. From experience, it does not work. And the research shows the motivation to change must come from within *ourselves*. We only do what we feel like doing. Wanting our horse to experience the development of resilience and build mental fortitude has to come from within *us*. Pushing others to change just makes us all push back.

The field of neuroscience explains why pressure does not work. We are wired to move towards what feels good right now, and to move away from what feels hard in the moment. We will always choose what feels pleasurable right now, avoiding discomfort. It is why equestrians struggle to offer agency (choice) in the other 23 hours of the day. What if the horse decides not today? Or worse, 'Not me?' Whilst we know what will cultivate a horse-human connected relationship (we help horses resolve negative experiences and offer positive experiences), the ability to respond to their needs, what your horse values, *before* our own, is hard.

Learning to separate our needs from the needs of our horse means equestrians say to themselves, 'This is going to be challenging, creating opportunities for my horse to have positive experiences, but I'm going to do it anyway.' We have to separate ourselves from the discomfort. We must decide to override the discomfort we feel as we unlearn and learn to help horses resolve their negative experiences (the purpose of conflict behaviour and therefore not that our horses are 'naughty,' but that they are trying to communicate that they need to behave in ways that yes, may be dangerous for us, in order for our horse to resolve their negative experience).

Dr Tali Sharot, a behavioural neuroscientist who integrates neuroscience, behavioural economics and psychology, believes in life. We tend to focus on our differences, because those carry the most amount of information about what makes each person unique. Our brains are organised in similar ways and will react similarly given the same situation. According to Sharot, the similarity of the circuitry in our brains may not be an easy notion to accept. From our point of view, from inside our skull, our mental world feels completely unique. It is difficult to imagine that people around us share very similar neural patterns of activity; similar mental states, thoughts, and feelings. How can this person outside of me be so much like me? Yet the basic architecture of the brain is remarkably similar. Everyone thinks they are the exception, and we are not the ones unintentionally harming horses. We are also the ones who believe we can 'fix' an unwanted behaviour, perhaps a functionally effective behaviour that the horse is highly motivated to repeat in an attempt to resolve a negative experience. Being passive-aggressive, constantly bringing up what needs to change for the horse or using threats as a way of trying to pressure someone else to change, will always backfire, according to neuroscientists. Brain scans show that when someone is telling us something negative, for example, your horse goes to the back of the stable when they first see you carrying the tack to avoid a bridle and saddle fit, meaning your horse is trying to resolve a negative experience, either (a) process of having the bridle/saddle fitted or (b) the subsequent consequence of fitted tack—being ridden, our brains literally tune out. The part of the brain that is listening to negative information turns off! This means all those worst-case scenarios, passive-aggressive comments, and eye rolls that equestrians do to each other are not even registering in the other's brain. You are wasting your time, words, and breath. Hence, so many equestrians are frustrated and stressed out by other equestrians. We need a different approach to enhancing horse welfare. We need a welfare horse sports mindset which, if we get right, may inspire equestrians to want to change their 'this is how I've always done it' mindset for themselves. Say you walk into the arena and your friend is using their leg constantly, making sighs, and being visually emotionally dysregulated, stating loudly enough for you to hear (or anyone for that matter), 'He is so lazy!' You are not going to be thinking about anything you just learned about the human brain and the science of motivation. Instead, you immediately feel annoyed. You are thinking about how easy it would be for your friend to train her horse to respond to her light leg aid by applying learning theory, specifically removal reinforcement (negative reinforcement). You even showed your friend a sensitisation process last weekend! You can't help yourself from audibly sighing, 'Is that right?'

Since you are coming from a place of judgement, not acceptance, you are not thinking about all those little steps she needs to start doing to help the horse resolve her relentless leg pressure. You are just thinking about how easy it is for your friend to repeat what you showed her at the weekend! Instead, the reality, the science, and the truth is that change is hard for everyone—and we jump right into frustration over the fact that your friend is not doing what you want her to do, despite being shown.

Our audible sigh of disgust is not going to motivate our friend to start correctly applying removal reinforcement. In fact, those sighs of contempt are going to double down on your friends belief that her horse is the exception and that he is 'just lazy.' No matter how caring your intentions are behind the audible sigh, your friend is feeling like you are trying to fix her incorrect application of removal reinforcement—which feels like even more discomfort, which means she is going to move away from you and you never do get another chance to enhance her horse's lived experience. You are not alone. Pressure doesn't create change—it creates resistance to it. I did this too. In my early years of applying learning theory, I coached clients in great depth in the correct application of removal reinforcement. At best, I would see a new client one more time. The pain (discomfort) of learning new skills (procedural knowledge) when clients shared they had been in the saddle since childhood felt harder than gifting their horse clarity of communication and positive mood behaviours. According to psychology research, growth has to feel uncomfortable; but when we choose to embrace discomfort, it can be motivating. When I changed how I started coaching, asking 'What did they want to achieve by the end of the session?' and actively listening to *their* framing of a challenge they were currently in, I could give the rider something they thought they needed (the most common response 'horse on the bit') by training stride lengthening (as a consequence of legs taking longer strides, head and neck lowers) as a result of changing the head and neck position, what the rider wanted, and I got to see the horse for more sessions. In the field of applied behaviour analysis, offering and providing what someone thinks they need in order to get an opportunity to provide positive experiences for the horse, the 'motivational grab,' gives riders control.

Because when our ability to control our environment is removed, people feel distressed and anxious. When we try to exert control over someone else's choices/behaviour, they instinctively resist our attempt to try to control them. Instead of inspiring change, our pressure creates a battlefield over control. It is why physical constraint is psychologically disturbing for humans and animals. Even infants prefer to exercise their ability to control their surroundings.

The fear of losing control accounts for some of our phobias and deep anxieties. It is why physical constraint is psychologically disturbing for humans and animals. Aerophobia is the fear of flying. Or is it? What we seem to dread is not what we fear. Neuroscientists state that acrophobia can be about control. Every time we fly, we hand over control to the pilot.

A welfare horse sports mindset requires overcoming our own instinct *for* control and recognising others' need for agency. When we pressure equestrians to update their prior knowledge or tell them how to manage their horses' lived experience, we are threatening the hardwired need for control over their decisions and actions. So, in order to influence equestrians and enhance horse welfare outcomes, we need to overcome our own instinct for control. Choosing influence over control involves behaviours that inspire equestrians to take ownership of their horses' resilience (fortitude), competence, and choice availability. When we stop trying to pressure equestrians to change and provide a safe place to be curious and make mistakes, build a culture of growth we embrace our positive influence. Like horses, people are social beings, highly influenced and inspired by those around us. This has been proven by decades of research on human behaviour. When we see something working positively for a horse owner, it often makes us feel curious too. When a rider is getting the results we want, having fun, or making a dressage test look pleasant for their horse, we are hardwired to move towards it. It's why, when we hear a friend raving about their new equestrian coach, we naturally want to go and watch their training. If you give your horse a social stable (head and neck interact with the horse next to them), our friends are more likely to take their stable bars down and provide social contact (positive horse welfare). Neuroscientists refer to human behaviour being contagious as a 'social contagion.' How we enhance horse welfare outcomes is through our own individual influence, modelling the behaviour we want for horses. When a professional dressage rider took to socials in 2025, with a video of her internal stable panels (with metal bars) being removed to witness elite dressage horses mutually groom one another, the video went viral. Through a change in the rider's *belief* about stable horses (updated prior knowledge with evidence-based knowledge), this rider accommodated new knowledge (social windows improve the lived experience of horses) and provided the opportunity for positive experiences. Having access to touch the neck and head of another horse or remove themselves (agency) is an example of modelling the change we want for our horses. How did an elite rider come to change her belief? Remember, a belief is an estimation of whether something is true or not. And beliefs are wired into the brain through experiences and reinforcement. In cognitive science, researchers discovered we process *new* knowledge in two ways. We either fit new knowledge into our existing framework, 'assimilation,' or we update our mental model (schema) because the new knowledge is too strong to ignore, shifting our estimation of the belief in something as true. Knowing how minds change (assimilation and accommodation) has meant I have strong beliefs, loosely held. It helps me move from curiosity to change because of new knowledge. In the words of Maya Angelou, 'do the best you can until you know better. Then when you know better, do better.' Remember, your mindset is your collection of attitudes (emotional positions toward specific ideas, people, or situations). We change our position towards an idea, person, or situation when we shift our estimation that something is true—our belief. Equestrians can change their behaviour, as with the dressage rider taking down the metal bars in order to give dressage horses social partners. When cognitive scientists found that the ability to change is what we used to think is dependent on our propensity for *intellectual humility*, a correlation between intellectual humility and having a reflective thinking style was uncovered. From a series of five studies involving 1189 participants, scientists examined how intellectual humility relates to acquiring new knowledge. Having intellectual humility requires a higher need for cognition and intellectual engagement, suggesting intellectual humility fosters authentic openness to new knowledge and learning. Individuals who can recognise their limitations of understanding and adapt their thinking are able to integrate new information and perspectives. Changing our *beliefs* (through assimilation and accommodation) changes human behaviour, enhancing horse welfare. A rider who provided access to social partners for the other 23 hours of the day did so because she changed her mind (belief); agency over proximity to another horse matters to the lived experience of horses. By accommodating new knowledge, she demonstrated the ability for intellectual humility, *applying* the Five Domains (2020) Model and removing the stable bars. From the evidence so far, horses accessing social partners is the heartland of positive experiences.

People tend to think of 'perspectives' as different views (vision) on the same thing. However, journalist Monica Guzman explains that perspective is the unique path people walk, resulting in our views/vision of the same thing. Our knowledge or understanding can go beyond our own 'horizon' or range of vision, and our perspective is to see what our *mind*, not just our eyes, sees. Notably, our range of vision is limited by our 'horizon,' and therefore what is beyond my horizon is beyond me. What I love about Guzman's explanation of perspective is that it draws our knowledge with obvious bounds and limitations. Our proximity or closeness to the horizon provides our perspective; we can't know what we are not close to. Being out on the water, in clear conditions, our sight is approximately 3 miles. When something is beyond my horizon (3 miles), it is beyond me. I don't see it, so I can't know it. So how then do we 'see beyond our horizon'? Trying to look beyond our own perspective means a struggle to see clearly. In fact, we don't see anything with certainty.

We can't prompt AI with 'I don't know what I don't know, can you tell me what I don't know?' However, we can get curious. Ask 'What am I missing?' According to Guzman, this is *the* question. How we relate to each other can be explained by Guzman's SOS model, which identifies three patterns of human behaviour in our equestrian communities; sorting, othering, and siloing.

Sorting is an invisible process where we end up close to people like us. It turns out, we like to be around people like us; people who share our beliefs (estimation something is true), attitudes (positions toward specific ideas, people, or situations), and values (where we spend our time and resources). 'Our people' make us smile, having the opposite effect of feeling anxious. Being around people like us feels familiar and comfortable. Even inside our chosen equestrian communities—livery yard, regional riding club pony club branch, or equestrian sport discipline—we soon find our people. We kick-start conversations with easy common ground, like the weather as an equestrian or our horse's sentience. The chitchat flows, probing for connection bright spots. When we know and like someone, conversations pick up where we left off, steering the horse chat to topics we know we find interesting and useful to us. We have the benefit of a shared path, earning trust, and practising how conversations between us make us feel.

When we don't know someone, conversations are all about discovery. Like horses searching for the positive experience of *mutual* grooming, equestrians search for other equestrians with mutual beliefs, attitudes, and values. According to Guzman, some people are too different to find common ground with quickly. Equestrians chatting with non-equestrians might prompt us to rapidly return to equestrians, people who we find easier to talk to. Finding and huddling with *our people* is what social scientists call 'sorting.' Sorting is what happens at equine conferences when everyone has found and sat with their favourite equestrians and hasn't moved seats for the two days of the conference. Conversations flow for reasons we feel, know, and appreciate. Being alike connects us, and social scientists call this love of sameness the 'birds of a feather' phenomenon. We bond more easily with people who are like us, receiving the growing recognition that strong relationships in community benefit health. We crave community and connection. Sorting makes it easy. Social platforms reward like attracting like, curating our online social lives and reinforcing more sorting. Our love of sameness keeps socials spinning with likes, recommended groups to join, and profiles to promote to our own equestrian voices. It feels so good to find our people that we have stopped spending time with equestrians who don't think the same way. Creating opportunities to develop disagreement and perhaps understanding new perspectives and practices necessitates friction in order for the brain to do 'accommodation,' the process of updating the mental model of an estimation on whether something is or is not true (belief).

The downside of the sharing feel-good from sorting is we end up sharing our blind spots too. Then our whole community has the same blind spots, such as interpreting horse behaviour. Judges might confirm each other's ignorance regarding the purpose for the horse of mouth opening, collectively rewarding repetitive and prolonged displays of competing motivations—tail swishing, mouth opening, etc. When we surround ourselves with people who reflect the same basic set of perspectives, we find it harder to grasp any others. Not because equestrians are incapable of grasping them, but because we are less likely to be given the chance.

When equestrians around us shape so much of our thinking, it stops being thinking. We stop colliding with equestrians who disagree with us, maybe because equestrians feel burnt out by the 'righting reflex' that plagues equestrian culture, leading to missed opportunities to see different angles or get a fresh view of a complex challenge, like sustaining horse sports. When this happens, it is a signpost that sorting is keeping us from seeing the future of equestrian sports clearly.

In researching this book, I came across research explaining that when we unite around what we have in common, we also push away from those not like us. *Othering*. Guzman clarifies othering as ranging from group boundaries, to hate and bullying in between. Riders Minds researched for Anti-Bullying Week in November 2023, revealing that 76.5% of equestrians have been bullied, of which 77.3% felt unable to speak out. A further 85.3% said that they have witnessed bullying within the industry, and 56% said they have seen others being bullied online. Fourty-two percent felt unsafe online.

Pragmatic innovations to our global equestrian challenges enhance horse welfare and look like small steps—providing horses with 'social windows'; practical, feasible, and solution-focused action to a real-world problem. This is a small *and* significant change for the horse. It is one solution in our current economic, technological, and social context that has rippled positive horse welfare.

We can't expect horses to experience clarity of communication when we clash our reins and legs together, or when we are tight on time and we choose to replace in-hand training with a short ride. We should model the behaviour we want for horses. Research explains that for a person to change behaviour, we first have to change our beliefs. We need to believe it was *our* idea. If every day, before we get on, we train in hand; go, stop, park, and head down at the mounting block, when someone on the yard sees us do that every day for a few weeks, our behaviour starts to influence others subconsciously. When equestrians see the relaxation of a horse at the mounting block, us smiling and in a better mood from our in-hand training, we ripple our positive emotions *because* of in-hand training. We see it happen again the next day; simultaneously, your friend's horse is avoidant and aroused at the mounting block, and your example influences friends, even though we are not consciously thinking about it. Some time later, walking past the arena, your friend is training head down, close to the mounting block.

What I love about modelling behaviour we wish to see in the world is that our friend is now in the arena enhancing her horse's learnt responses in-hand, thinking it was her idea. The ripples of a welfare horse sport mindset at work are magic. We were just going into the arena, testing our horse's in-hand responses to go, stop, immobility ('park'), and head down, enjoying attuning to how our horse feels about our interactions/training.

We can use science (neuroscience) to inspire any equestrian in our lives to change:

- Accept that we can't control another's behaviour or actions. People only do what they feel like doing. The welfare horse sports mindset accepts equestrians for who they are and where they are on their own equestrian journey. Our mindset meets equestrians where they are.
- Model change we want for horses.
- Make it look fun.

Research shows our influence is highly effective but requires a lot of patience because it takes time for our influence to take effect in another's brain. The key to influencing change for horses is to focus on our own behaviour and emotions about it. Have no expectation that equestrians close to us will change, no matter how hard it is not to go in and 'rescue' our friend's horse (increased arousal and negative feelings at the mounting block, due to the association of the mounting block with relentless rider pressure during training). Known as the 'backfire effect' bias, the more we show up and 'offer help,' the opposite happens. As in a 'backfire,' we start wondering how is our friend doing? Then it dawns on us that they are now at the yard on a different schedule, training in the arena at different times. Expecting others to change (because their horse will be safer, more competent, and more resilient) will lead us to start resenting our friends when they don't. Focus on our horses, model the behaviour we want for all horses, and our attitude towards applying learning theory, because it makes *us* feel good. Let the science of influence do its work. Be prepared to give it 6 months or more. Give your horse clarity of communication, scratches at the base of the wither, space between the nasal plane and noseband, and even choice to be trained. (In the arena, the horse chooses to present at the mounting block in order to be reinforced [scratched/food], followed by you mounting, or not [there are small approximations in your training before choosing to be ridden; the horse is choosing to be reinforced, or not].)

What if we can't wait that long to enhance the mental and/or physical well-being of the horse?

Apply the advanced science of influence.

Motivational interviewing (MI):

- Ask open-ended questions
- Back off and observe their behaviour
- Celebrate all/any progress (especially the smallest steps)
- Model the change we wish for horses

The conversation (asking open-ended questions) needs to be in person, without any alcohol involved, or time pressure to finish quickly. Do not allocate only 20 minutes on the phone, or whilst standing at the water tap waiting to hose your horse's legs. This is a conversation where our welfare horse sports mindset gets to practise curiosity and compassion. It is not an invitation to complain or rant about how frustrated and worried you are for their horse, your friend, or both. We are not there to be 'right.' We are practising communicating to neutralise tensions and create *space* for positive change to happen. Active listening then requires us to be present wholeheartedly and to practise not interrupting. I work on my active listening every day. Some days with more success, but I noticed as I practiced being present and actively listening, it does get more fluent. Never 'easy,' just more practised. MI requires active listening and the question, 'why?' The founder of the Toyota companies is attributed with the Five Whys Method, which helps engineers uncover the root cause of a particular problem and is now taught in business schools and engineering programmes globally. Each 'why?' question is about unlayering what your friend thinks the problem is and why they believe the horse is doing the dangerous/unwanted behaviour. Five whys later, it might be clear that the behaviour of the horse is terrifying the owner, and listening to why your friend is so terrified might illuminate a path of positive change. Getting to the root cause is nearly always (having done this exercise with hundreds of equestrians) a version of not feeling 'enough' or not 'good enough' for their horse, keeping your friend stuck in a version of 'doing what they have always done' because she is embarrassed or she is scared of getting injured, resulting in time off work, a drop in financial resources, and partner pressure to rehome her 'best friend.' Allow ourselves to listen without judgement. Acknowledge our need for control and drop the urge to fix. Suspend your 'righting reflex,' and as this comes up for you, ask yourself what is driving your need to be fixing for your friend and her horse. MI is effective in helping us influence an equestrian's motivation to change. Your friend talking about how *they* feel encourages them to think about the gap between what they want and their current behaviour. Instead of knowing what someone else should do, we are asking open-ended questions because we are curious about where equestrians stand with their horse's behaviour, and their own in response to this. Really listen, lean in, and try to learn about how an equestrian feels about the issue. Reflect back the answers they give. Whatever the answer is, we are not going to share our feelings about it. We are just going to repeat back their answer:

> So it sounds like…
> So it sounds like you are feeling okay with your horse at the mounting block?
> What makes you feel okay?

Continue listening with curiosity and acceptance and respond only with open-ended questions that repeat back what they have just said.

> How do you feel about that?
>
> What feels hard about it?
> Can you tell me a little bit about how long you've felt this way?
> So I hear you saying that you don't need me to do anything?

Our opinions are not for this conversation. As soon as we offer a welfare opinion, we will be pressuring the person, and it ends the effectiveness of MI.

What is so effective is people admitting to themselves there is a disconnect between what they want for their horse and what their behaviour is, impacting their horse's behaviour. Recognising for ourselves the tension between these two things is the point of MI. To create discomfort we feel internally. That tension is critical as it becomes the source of having the motivation to change. Seeing the disconnect between what they want and their current actions is what pushes them to eventually change for themselves. We can't insert ourselves; I know that's hard. But as horse welfare advocates, when we insert ourselves, we motivate equestrians to do more of what is not progressing positive horse welfare. This type of conversation is critical for awakening our motivation to change.

Now, back off.

The next step is to stop pressuring them. Don't expect equestrians to launch into action for the sake of the horse. We all need to marinate on it. This is why it is critical to model the change and make it look fun, giving them the freedom to figure out for themselves why this issue matters to their horse with all the positive horse welfare benefits.

We have to give our friends time. Eventually, their tension transforms into the motivation to be the change. After asking open-ended questions, observe the behaviour and stop.

One way to express control is to make a choice. Researchers found that choosing and having choice is inherently rewarding. Choice is also referred by animal welfare scientists as *agency*. Agency is the capacity of the individual to engage in choice when performing goal-oriented behaviours. Agentic experiences (the choice to engage in behaviour) infer from these agentic experiences based on available knowledge about the animal's motivation for engaging in the behaviour. Competence-building agency can be used to evaluate the potential for positive welfare and is represented by the Behavioural Interactions domain of the Five Domains. In 2020, the Model was updated to, amongst other things, include consideration of human–animal interactions. The most important aspect of this update was the renaming of Domain 4 from 'Behaviour' to 'Behavioural Interactions,' and the additional detail was added to allow this domain's purpose to be clearly understood as representing an animal's opportunities to exercise *agency*. We illustrate how the behavioural interactions domain is being applied to assess competence and resilience through the development of agency and the experience of positive mental states.

Discussing how we train horses has become contentious. Choosing curiosity over control (horses and people) is the heartland of the welfare horse sports mindset. Asking 'What am I missing?' opens possibilities including conversations where the outcome is not win–lose, I am right, you are wrong, but opportunities to think 'I never thought of it that way.' Curiosity, then, is a serious thing. As equestrian communities get more polar in how horses are trained, we are overdue getting serious about curiosity. What we learn relies on how open we are to learning and what room we make in our minds and conversations. If we treat curiosity as a passive state that turns on and off randomly, we miss opportunities to direct it and learn, asking what are we missing? No one field of knowledge, equestrian discipline, or practitioner/coach/trainer has 'all the answers' for you to have a 'better connection' with your horse. If they say they have, get curious. What do you need to know exactly? In listening to the answer, scan for humility. Curiosity flips on when we see a gap, any gap, between what we know and what we want to know. The absence of humility has equestrians knowing everything. Worse, coupled with a 'this is how we've always done it' mindset, curiosity

THE WELFARE HORSE SPORTS MINDSET

is killed. Curiosity at its core is an urge to close a gap in our knowledge. Researchers in the field of curiosity have created a framework called 'information-gap theory':

- *Notice the gap in your knowledge.* You know your horse is not feeling optimal about your coach asking you to increase the pressure down the reins. Why is this? What don't you know?
- *You want to close the gap.* You ask your coach why does your horse open his mouth in response to your stronger pressures? The coach replies he is motivated by the stronger pressure (aversive) to find ways (such as mouth opening) to resolve the strong pressure. Maybe the coach is now tightening the noseband in a bid to stop/deter said mouth opening.
- *The closer you feel you are to closing the gap, the more intense your curiosity will get.* Asking why, when the rein pressure increases, does she tighten the noseband so the horse can no longer resolve for himself the mouth discomfort? To get curious means asking what you see. If your coach fills your knowledge gap with a question such as 'Can you feel how much lighter down the rein he is now, because *we* tightened his noseband?'—get *more* curious. Why is a coach treating the symptom of stronger rein pressure and mouth opening with an action that creates more discomfort (crushing the soft tissue of the nasal plane), tightening the noseband? Identify a knowledge gap. Stronger pressure should only last momentarily, the duration to motivate the desired response. (Pressure motivates a behaviour; it is your release of the pressure that trains your horse.) If you are being instructed to increase rein pressure and *not* release for a desired behaviour such as slowing/stop, you are breaching First Principle 6: Regard for Operant Conditioning (removal reinforcement) to give your horse controllability of their world (leading to competency). How does your horse get to control the pressure down the reins? If your horse is communicating via mouth opening, your rein pressure is too strong. Do you know if you consistently release the rein *for* the correct response (deceleration)? If not, it is more than likely he is resolving your stronger pressures by opening his mouth. Opening the mouth is functional to your horse's lived experience. By being

physically able to open the mouth, he is resolving the experience of discomfort delivered down to his mouth via your reins. Going to tack/equipment to solve the escalation of pressure is akin to treating the symptoms, not getting closer to the cause. When we deploy curiosity, we can actually start the journey of enhancing horse welfare. Applying evidence-based knowledge enhances the lived experience of rein pressures for horses. If we are going to the noseband to solve the elevated rein pressure, we are creating conflict (confusion) in the life of the horse, explaining why your horse does not feel optimal about stronger rein pressure.

- *The more pleasure we get from learning, the less it matters if the learning is hard.* The ability to see the discourse between being coached to increase rein pressure on a sentient animal and the inability to recognise a horse's sentience and his need for controllability of his world (via removal of pressures).

Set your mind on the gap between what you know and what you don't. Set your mind on the hunt for new knowledge and you can do the first part of welfare horse sports, knowing better to then *do* better, enhancing the lives of horses. When we make assumptions about knowledge, including opinions, we are missing what is actually going on for horses. Assuming we choose our opinions is not what happens. According to the research, our beliefs (estimations of whether something is true or not) are formed over the course of our equestrian lives. This makes me curious; if we don't choose our opinions (they are the sum total of our beliefs shaped by our own lived experiences of equestrian culture), and our beliefs form naturally over the course of our equestrian lives, then equestrian culture (shifting what is currently normalised) is the heartland for positive welfare change. The belief that we can use horses in sport currently, with no attention on the role of the horse competing and their sentience (lived experience) being seen and valued, helps us understand how equestrians arrived at now doing horse sports *with* welfare. Normalised behaviour becomes the culture. Witnessing the systemic brutalisation of horses ('give her a smack') forms the belief (estimation of whether something is true or not) that horses should be dominated (dominance theory). When systemic abuse is normalised, we see children instructed to whip horses. Our opinions on horse training are not then shaped by current knowledge (freely available), that is, First Principles (ISES), but by our beliefs, formed over the course of our lives.

A welfare horse sports mindset asks:

1. *'Am I willing to believe that I am wrong about something?'*

- Is it safe to ask am I wrong about something? I think so. Because we have all been wrong about so many things before. But there is a problem. We don't know what we are wrong *about*. Observing that I am wrong, but I don't yet know *what about*, cultivates some humility and a willingness to be curious.

2. *'Which do I value more: The truth or my own beliefs?'*

- If I value my own beliefs more than the truth, I am going to defend myself, especially if my identity—who I am, what I do—is attached my beliefs. Why then would I be open to exploring curiosity or be willing to listen to current knowledge? To do welfare horse sports, you have to value truth more than your own beliefs, and you have to choose to experience your own intellectual humility.

With no more than these two questions, we can help our minds move from certainty to uncertainty, finding gaps that cultivate curiosity. If, that is, we are not preoccupied with winning. Focusing on our beliefs and opinions makes wanting to win a natural instinct, as does craving dopamine lollipops that satisfy us with the feeling of being right, rather than needing evidence that we have actually persuaded another equestrian to apply the science of how horses learn. When someone chooses to stand down in an argument over horse training, was anything learned? Was anyone persuaded? If the conversation or online post is hostile and combative, who would admit to being moved by it? How would you ever know? What feels like your victory might just be the other person losing steam, or you being better at on-the-spot communication. If we are truly curious to learn and get closer to the truth about horses, we need to build a mindset of helping equestrians understand each other. From my own experiences, you can't enhance horse welfare with a mindset of trying to outmanoeuvre or defeat discourse. If we are serious about staying curious with people who train horses differently from us, it is better to try not to win. If you do end up influencing someone's thinking about their horse's lived experience because of your conversation about the science of how horses learn, great. It is also unlikely to happen in a huge way right then. Instead, it will expand our certainty and rush to judgement for some acknowledgement of our rightness.

We believe in opinions to be fixed, but the welfare horse sports mindset recognises that our opinions are *curious conversations* waiting to be had with other equestrians, to enhance horse welfare outcomes. As equitation scientists explore more questions, our knowledge of horse training evolves. So should our beliefs. That is why we need to keep revisiting our equestrian norms (not the 'this is how we have always done it' mindset) and stop missing the mark in bringing in horse sentience into horse training and sport (see Appendix I). We need our beliefs to serve the welfare of the horse. We should ask ourselves: What concerns me? What gives me hope? We must not defend our current set of beliefs to each other, at all costs, at all times, but instead strive to represent horse welfare in ongoing conversations, placing the role of the horse at the centre of horse sports. Curiosity requires uncertainty, and uncertainty requires flexibility. If truth matters more than our beliefs, then we can choose to enter *bridging conversations* holding our beliefs more loosely. It takes courage to let our beliefs breathe in a conversation. Let gaps appear, without freaking out that we don't know everything about everything (control), for our horse. We can then explore our current beliefs, not to set out to prove, but to learn something. If we were less afraid to hold our beliefs loosely, could we consider new knowledge more freely?

The welfare horse sports mindset knows we don't choose our current beliefs. We show up attuning to the lived experience *of the horse*, holding our current beliefs tight enough to claim them, but loose enough to change them, by staying curious. Our beliefs about how to train horses are the natural result of our life's interactions, so trying to change our friends' deeply held beliefs about how they train horses is not dependent on how persuasive you are, but on understanding how minds change. Being curious when we talk to equestrians who think differently means not trying to win or change minds. Listening to what equestrians believe and how they've come to believe it requires exploring three things; the experiences that inform their perspective, the values that underlie their hopes and concerns for horse welfare, and the connections that hold them to their beliefs. Guzman refers to this as attachments. Seeing each other's attachments helps make sense of our beliefs. Where do you place your time, effort, and money (values)? According to research, it is not which attachments each of us hold, but their order of importance. The order of our attachments offers more insight than our commitment to anyone. When we make the assumption, 'if we are not motivated to enhance horse welfare, you must be against it,' curiosity is blocked. By assuming equestrians who disagree with us don't value what we value, we can instead ask, *What do they value more*? If we all want to solve our problem of using horses in sport ethically, then valuing having opportunities to actively pursue species-specific behaviours and individual capabilities for horses to thrive might be in a different order to your friend, because she walks a different equestrian path, aligning with different horse training beliefs. Becoming a detective and asking your friend, 'What are your concerns?' is another way for your friend to show you what she values. Researchers call this 'concern gathering,' the first steps in exploring sticking points in tough social issues. When researchers start with this question, they start where people really are, not where experts or organisations think they are. Taking into account equestrians' various starting points and concerns gets to the real values of equestrians, not our assumptions of what others value.

Attachments are anything that causes us to put pressure on ourselves, meeting a set of expectations about what our beliefs ought to be. Attachments can be positive or negative, helping us align closer to our values or making it extra hard to rethink and consider with fresh eyes, looking at evidence-based knowledge. Attachments can even make it hard to think for ourselves and ask, what is in it for the horse? As long as we have lives, we have attachments. Identities and positions give us a sense of belonging that makes life good, and some spaces/places where it gets harder to get curious. How do we hold our beliefs tight enough to claim them *and* loose enough to change them?

Cleaning up the digital apps we have on our smartphones offers a great visual of holding our beliefs tightly, until new knowledge comes along. Your smartphone operating system starts to shake when we want to rearrange them; delete, move, or regroup. We want to do with our horse training beliefs what our phone does to our apps. Exploring new knowledge, I don't want to lock down in certainty even the beliefs that stem from who I have become today. I want to stay curious. I want them to shake. I know the First Principles of Horse Training (ISES), making it my life's work applying the ten principles; both in hand and under saddle. However, these principles have seen three iterations just in a short space of time. Each time I loosened my beliefs of what I thought I knew, pressing 'update' on being notified, there is a latest version (like when our phone notifies us of a new update). How many of us see a notification on our smartphones sharing an update is ready for us, only to then do nothing? We reason that the short time for the phone to be turned off and receive an update is not right now. And so updating gets delayed or stays in its existing, now outdated version. The problem with no action on notification is your phone needs the update. The latest enhancement is to improve its functionality, speeding up and fixing the 'bugs' from previous versions. When we forget or 'tap out' of updating our equestrian beliefs, deciding not to accommodate new knowledge and stronger evidence, either because they shake our existing beliefs or we think we don't have time, we are standing still. Whilst research keeps progressing what we know, even seeing the updates and knowing better, does not mean for horses, we *do* better. Progress for enhancing horses' lives comes from choosing to have our beliefs shaken. Like adopting a growth mindset, we cannot access growth when we choose comfort. Only when we step into being in discomfort, small discomforts such as actually seeing your horse's facial expressions, followed by reflection, do we choose to get curious. Owning what just happened for your horse and understanding that how she feels may be in response *to* me, is uncomfortable. Responding to our horses and being responsive to their affect (emotional state) is one of ten First Principles (ISES) (more on these later) and an important indicator of your ethicality. Asking if our interactions with horses are ethical is the first step, according to researcher Dr Rosalie Jones McVey in her latest book, *Horse–Human Relations and the Ethics of Knowing*. There are two ways we can know we are ethical in our horse training; responsivity *to* horses *and* reflection. Responding to the horse starts with first *seeing* our horses; *their* lived experience. How do our beliefs keep our eyes closed to how our horses actually feel? Knowing better is not the same as *doing* better. The 1999 film, *The Matrix*, is a story set in the future where humans are grown by sentient machines and kept imprisoned in 'the Matrix,' a virtual, computer-generated world. Our beliefs are like the choice in *The Matrix*. Choose the blue pill and you continue to normalise not seeing the actual lived experience of horses. Choose the red pill and you choose the truth. The film's hero, Neo, decides to know the truth. He was living with the discomfort of feeling like 'a glitch.' Choosing to scan for what is happening to our horses' mental and physical state is like being in an equestrian matrix, having swallowed the red pill. Faced with the decision to get closer to the truth about horses or double down on our existing beliefs (shaped from a time we did not have access to new knowledge), choosing to take the 'red pill' matters to horses (see Figure 2.2).

A welfare horse sports mindset holds existing beliefs loose enough to *see* the lived experience of the horse; noticing ears widening at the base, 'worry wrinkles' above the eyes (can also become triangulated), tense lips, squared-off muzzle, and squared dilated nostrils. It is not limited to noticing these changes in your horse, but noticing that your horse is communicating with you constantly. The subtle posture changes during routine interactions, or the less subtle look away as you approach with your saddle and bridle. Our existing beliefs, shaped by our past experiences, can keep our eyes closed. By letting my beliefs 'shake' like apps on a smartphone, I can move them. I might even delete them altogether. Or I might promote an app to my home screen and bump something else out of the way. By having beliefs that 'shake,' it gets easier to stay flexible, curious, and agile about equestrians' beliefs, seeing how their beliefs shake too. When this happens, evidence-based knowledge is readily shared from one to another, editing beliefs we used to

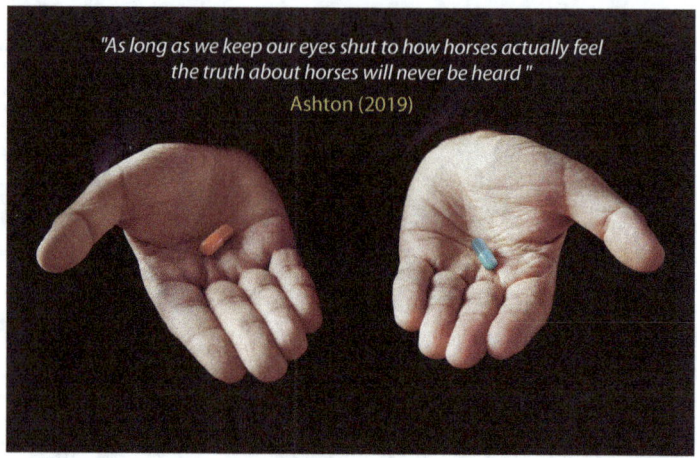

Figure 2.2 Our Beliefs Are Like the Choice in *The Matrix*—Choose the Blue Pill and You Continue to Normalise Blindness to the Reality of the Lived Experience of Horses. Choose the Red Pill and You Choose the Truth.

hold, and updating our beliefs, which can literally transform the lived experience of horses.

 REFLEXIVE PRAXIS

Have you swallowed the 'red pill'?

How does seeing the truth about how horses feel (pain, face conflict behaviour) make you feel?

COMPASS GUIDELINES FRAMEWORK

Misinformation in equestrian communities is not just about the facts—it's also about trust, identity, and tradition. When we hear a confident coach say 'this is how it's always been done', that statement can carry more weight socially than the quiet voice of new evidence. This is exactly where the lastest *COMPASS Guidelines* offer more than a research tool—they provide an equestrian literacy framework (McGreevy et al., forthcoming). Each element of COMPASS acts as a safeguard against misinformation:

- **C—Controls and Calibration**: Protects against biased measures that create false 'evidence'.
- **O—Objectivity and Open Data**: Transparency ensures that knowledge can be checked, not hidden within closed circles.
- **M—Motivation and Methods**: Keeps learning theory and behavioural science at the core, instead of heritage traditions.
- **P—Precautions and Protocols**: Applies the precautionary principle when evidence is unclear (what I call *mid-information*).
- **A—Animal-Centred Assessment**: Demands we see the horse's subjective experience, countering human-centred myths.
- **S—Study Ethics and Standards**: Ensures independence from commercial or competitive bias that often fuels disinformation.
- **S—Species Relevance and Scientific Rigour**: Prevents over-generalisation and ensures practices fit equi-centric knowledge, not human convenience. COMPASS does more than guide science. It helps all of us—trainers, coaches, riders, and researchers—resist the pull of identity-protecting misguidance and heritage practices. In short, COMPASS is a framework to *practise* intellectual humility. When misinformation thrives, horses suffer. Implementation of the COMPASS framework is benefits horses.

TOOLS FOR TRANSFORMATION
FRAMEWORKS AND MODELS

Misguided—Misinformation. When Falsehoods Feel True. According to the American Psychological Association, *misinformation* is false or inaccurate information, while *disinformation* is false information deliberately intended to mislead. In media literacy, the term *misguided misinformation* has emerged to capture the blend of both intentional and unintentional falsehoods. In equestrian culture, these concepts are not abstract—they are lived daily in our stables, arenas, and competitions. Consider the widespread belief that scraping water off a hot horse speeds cooling. Research shows the opposite: leaving water on increases cooling efficiency by the horse. Equestrians still scraping water off in 2025 are not uninformed—they hold a belief. But it is a misinformed belief, sustained by tradition than by evidence.

The Many Faces of Misinformation. Misinformation is an umbrella term with several layers: *Misinformation*—untrue beliefs shared without ill intent. *Disinformation*—deliberately misleading claims, spread to achieve a goal. *Mid-information*—when the truth is not yet fully known. An equestrian example of mid-information is the question of how horses experience the curb bit. Riders report that the curb rein 'feels lighter' than the snaffle bit (bridoon) yet it is currently unknown why? Greater pain for the horse? Resulting in a lighter rein pressure reported by riders. How the horse experiences the curb bit remains under investigation. When evidence is incomplete, equestrians operate in a grey zone, filling the gap with tradition, assumption, or wishful thinking.

As Facciani (2025) reminds us, misinformation and mid-information are not new. They have always been part of human culture. What matters is how equestrians navigate them: with humility or defensiveness

Misguided, Not Stupid: Identity and Belonging. Equestrians who hold onto falsehoods—such as the 'naughty horse' narrative—are not unintelligent. For the most part, equestrians are *misguided*. To be misguided is to be shaped by social forces, not simply by a lack of facts. Facciani (2025) highlights humans are deeply motivated to protect our identities. For equestrians, identities such as 'professional coach,' 'competitor,' or 'traditional horseman' carry social value. These identities bond us to our communities, providing self-esteem, belonging, and meaning. When a belief—no matter how misinformed—is tied to an identity, letting go of that belief can feel like letting go of the community itself. Thus, intelligent, thoughtful riders may maintain misinformed practices, because the alternative is social isolation.

WHY TRUST MATTERS MORE THAN FACTS

The *information deficit model* suggests that if people simply had access to the same research or attended the same conferences, they would change their behaviour to enhance horse welfare. But media literacy research shows this is naïve. Belief change depends less on exposure to facts and more on *trust in the source*.Trust is not only about competence but also about *warmth*. Equestrians are more likely to believe a coach who they feel cares about them and their horse than a stranger citing journal articles. This is why welfare-based knowledge delivered with empathy, as well as expertise, warmth as well as competence, are powerful predictors of trust.

Rational Falsehoods: Why Some Beliefs Persist. At first glance, it may seem irrational for a rider to continue patting their horse as a reward when evidence demonstrates horses prefer wither scratches (allogroom each other). But from a social psychology perspective, it is rational. If scratching instead of patting leads to rejection by a coach, or friends, rider's risk losing friendships and social support.In this light, the persistence of misinformed beliefs is understandable. Beliefs are not just about horses; they are about *belonging*. To protect identities and relationships, riders may uphold traditions even when they conflict with evidence.

Equipment as Identity: The Double Bridle Example. Equipment in equestrian sport often functions as an identity marker. In dressage, the double bridle (bridoon bit and curb bit together) signals to dressage culture refinement, progress, and competence. Judges reward correct use. Riders feel validated and the equipment becomes part of a feedback loop of self-esteem. To ride without a double bridle is a technical choice and an identity statement. The opposite identity—riding bridleless or with less headwear equipment (including metal in the mouth)—carries its own meanings. Both are sustained by skill and community validation. At the same time, photographs of horses showing blue tongues from rider rein pressure wearing a double bridle have brought horse welfare into plain sight. Welfare is about what head control equipment *means*—both in terms of the horse's lived experience (affect) and in the social world of equestions.

Amplification: Social Media and Echo Chambers. Identity bias does not exist in a vacuum. Social media platforms magnify it through sorting, othering, and siloing (SOS). Riders see curated feeds of equestrians who think like them, reinforcing existing beliefs and identities. This "hyper-echo chamber" effect amplifies misinformation. Identity becomes reinforced through constant affirmation, and alternative perspectives are filtered out. Social identity complexity—the fact that each of us holds multiple, intersecting identities—offers some protection. A rider who is also a welfare advocate, a scientist, or part of multiple equestrian subcultures may have more flexibility in updating beliefs. But when identity is narrow and reinforced online, susceptibility to misinformation grows. Equestrians are most vulnerable to misinformation when our identities are strongly tied to tradition or heritage. They exist in homogenous networks that reinforce existing beliefs. The social costs of change outweigh personal convictions. Midinformation creates uncertainty, and tradition can rush in to fill the void. The antidote lies in diversifying identities and communities. A welfare horse sports mindset asks riders to seek different voices, different contexts, and different ways of being equestrian.

Misinformation in horse training is not just a problem of facts. It is a problem of identity, belonging, and social reinforcement. Riders are not irrational in maintaining misinformed practices—equestrians are rational within our social worlds. To shift these worlds, equestrian communities must embrace diversity of thought and intellectual humility. This sets the stage for the frameworks that follow. The *COMPASS Guidelines* offer not only a research standard but also a literacy tool to resist misinformation. Each letter of COMPASS—Controls, Objectivity, Motivation, Precautions, Animal-centred assessment, Study ethics, and Species relevance—functions as a safeguard. When misinformation thrives, horses suffer. When COMPASS framework is applied, horses benefit.

A welfare horse sports mindset centres on the principle of leaving horses better. This involves not only utilising the Five Domains Welfare Framework, which culminates in the assessment of the horse's mental state as the sum of the other four domains, but also actively navigating daily interactions with this principle in mind. For instance, consider a riding lesson where a coach advises increasing rein pressure. A rider might comply but then sense that they are 'holding' the horse, experiencing their own physical discomfort and feeling that this level of pressure is not 'right' for the horse. An ethical response, rooted in a welfare mindset, involves attending to the horse's reaction to this pressure, such as opening their mouth. The example further illustrates a situation where a coach might counter this by tightening the noseband. Cultivating this welfare-focused approach begins with awareness through diligent observation, then allowing curiosity to guide our understanding. This practice extends to spectating, such as during a clinic, where one might notice a horse displaying signs of a negative experience in response to the instructions of the coach. These signs can include changes in facial expression and subtle or more obvious behavioural cues like tail swishing, triangulated eyes, and a wide base of the ears' position. Such observations challenge existing beliefs about what is 'right' in equestrian practices. The concept of leaving horses better necessitates a critical examination of whose perspective of 'right' is being considered. Common terms in equestrian education like 'correct' and 'good' should be continuously evaluated against the horse's lived experience. Therefore, welfare horse sports requires a collective commitment to asking 'How?' This question normalises a deeper investigation into the horse's experience, moving beyond surface-level observations and seeking to understand the underlying reasons for their responses. While using common equestrian terminology might seem like a shortcut that saves energy, the welfare horse sports mindset prioritises this deeper understanding and ability to think critically about what is happening for the horse. I like to refer to the ability to think critically as 'getting under the hood' of what is happening for the horse. This question embodies what it is to think as a realist, bringing to welfaer horse sports a realist informed lens.

PRINCIPLE OVER METHOD

The concept of 'horse justice' often evokes thoughts of legal systems rather than the fundamental responsibility to safeguard the lived experience of horses; this prioritises legalities over what is inherently right for the animal. However, identifying the shift from horse sports *with* welfare to welfare horse sports, which embodies a equi-centric-centric mindset, involves building trust through consistent, small actions aimed at identifying and resolving negative experiences and providing positive ones. This process should be applied to each of the four physical domains of the Five Domains Welfare Framework, with the fifth domain, the horse's mental state, being the culmination of these other domains. Demonstrating ethical responsiveness to the horse is not only inherently right but also essential for collectively earning public trust and ensuring the sustainability of horses in sport. Cultivating this welfare horse sports mindset begins with keen observation and a curious mind, normalising a deeper investigation into the horse's experience by asking 'How?' This question encourages us to move beyond surface-level understanding and delve into the reasons behind a horse's responses, challenging existing beliefs and continuously evaluating common equestrian terms like 'correct' and 'good' against the horse's lived experience. The regular practice of applying the First Principles of Horse Training (ISES) offers focus and intention on building equine competence, capability, and resilience (see Figure 3.1). Ultimately, the core principle of a welfare horse sports mindset is leaving horses better through our interactions.

TOOLS FOR TRANSFORMATION: FRAMEWORKS AND MODELS 63

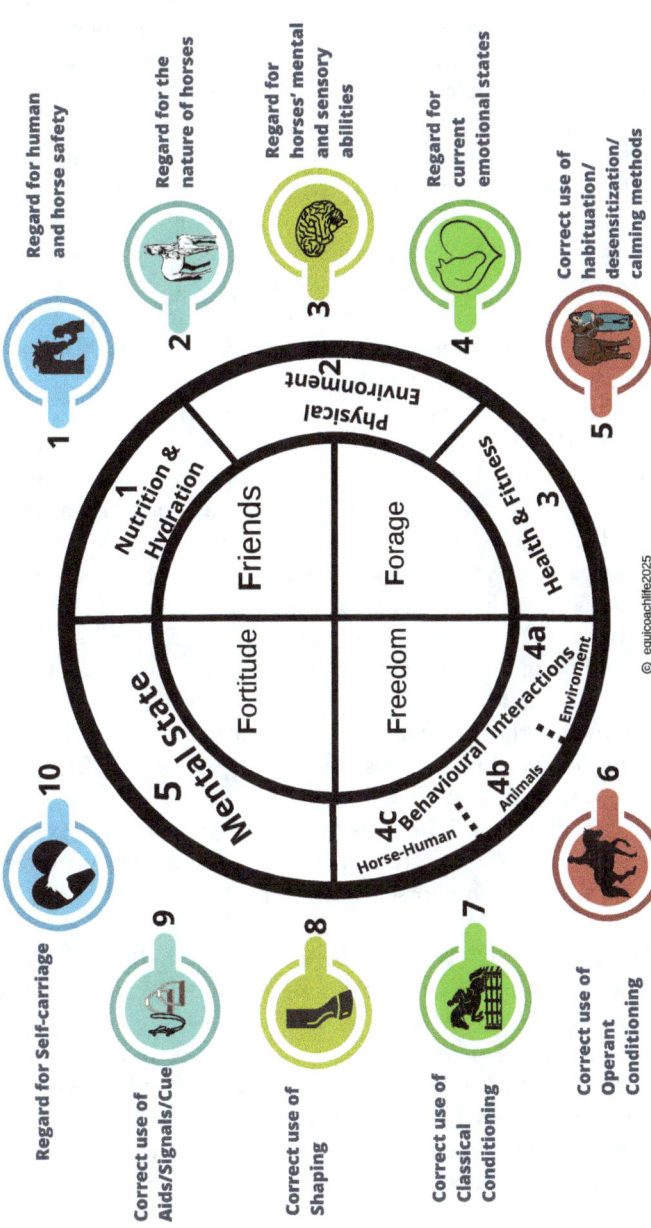

Figure 3.1 The Three Layers of Welfare Horse Sports

First Principles represent foundational knowledge and understanding, serving as the most fundamental and irreducible truths that can't be broken down further. This type of thinking involves deconstructing complex ideas to these core elements in order to build solutions from the ground up, and it is applied across various fields. In the context of Welfare Horse Sports, the application of evidence-based knowledge is guided by these First Principles. Specifically, the First Principles developed by the ISES (2018) provide a framework of peer-reviewed and up-to-date procedural knowledge, with the understanding that these principles evolve as research advances. These First Principles are not proprietary to any single coach, trainer, or organisation but offer a trusted foundation for equestrians to develop competence and capability in horses. In an era saturated with information and misinformation (the infodemic era), the ISES First Principles act as a reliable guide, much like a lighthouse for ships, steering equestrians towards ethical knowing. Comprising ten principles that span all disciplines and stages of horse training, they recognise that every interaction with a horse is a learning opportunity, whether intended or not. From foals to high-performance athletes, the application of these First Principles aims to provide horses with predictability and controllability over their environment, thereby enhancing competence and resilience, promoting predominantly positive mental states. This aligns with the core of a welfare horse sports mindset, which emphasises leaving horses better. As this mindset encourages us to ask 'How?' to understand the horse's experience, the First Principles provide a scientifically grounded framework for answering this question and ensuring our interactions contribute to the horse's well-being.

Like a fine perfume, Welfare Horse Sports unfolds in three layers. The top notes—the four Fs: friends, forage, freedom, and fortitude—represent the mental state of the horse, sitting under Domain 5 of the Five Domains. The heart notes—the Five Domains themselves—form the core identity of welfare assessment and have been adopted across global equine sport governance. The base notes—the ten First Principles (ISES)—anchor horse-centric practices (different methods) with an evidence-based foundation.

Together, these layers develop over time, just like the evolution of an ethical, welfare-centred equestrian mindset.

Each of the First Principles is covered in detail (application) in McLean (2024) *Modern Horse Training Volume 2: Principles in Practice* and can be seen on the ISES website: *www.equitationscience.com*.

Types of Scholarship (Boyer, 1990):

- *Scholarship of Discovery*: Conducting research to advance knowledge
- *Scholarship of Integration*: Synthesising information across disciplines
- *Scholarship of Application*: Applying research findings to solve societal issues
- *Scholarship of Teaching*: Investigating teaching and learning processes

Applying the First Principles (ISES, 2018) to enhance the lived experiences of horses has led to collaborations within various equestrian communities, including organisations like Blue Cross, Redwings, the Horse Trust, Bransby, UK mounted police officers, Mount St John, UK Equine Veterinary Practices, the British Veterinary Nursing Association (BVNA), EquiPilates founder/community, an Italian professional show jumper, and the Association of British Riding Schools. This work of translating research into practical application to address real horse training problems is characteristic of an applied scholar. Also referred in academic communities as 'pracademic'. Volpe & Chandler (2001) are credited with coining pracademic and popularising the term in academic literature. This approach intersects the application and integration of scholarship, aligning with Ernest Boyer's 1990 framework that recognises diverse scholarly activities. As an applied scholar focused on resolving horse training challenges for horses, equestrians, and governing bodies (Association of British Riding Schools, Equestrian Centres and Livery Yards [ABRS+]), the integration of the First Principles and the Five Domains (2020) Framework has been instrumental in exploring solutions for unwanted equine behaviours. This involves engaging with scientists, coaches, and policymakers to ensure that equitation science research is grounded in real-world contexts. A concrete example of this pragmatic innovation in sector-wide application of the First Principles (ISES) was the facilitation of an online Community of Practice (CoP)—Espresso with Equicoach.

 REFLEXIVE PRAXIS

For each of the First Principles (1–10), explore the reflexive questions:

1) In what ways does my own comfort with risk influence the safety of horses and riders I coach?
2) Do I accurately assess the cognitive and perceptual limitations of the horse, avoiding anthropomorphism or underestimation?
3) Do I actively avoid methods based on dominance theory, and how do I ensure my approach supports the horse's natural behaviours and well-being?
4a) How do I evaluate and respond to the horse's emotional state to ensure they remain within a low-arousal, learning-friendly zone?
4b) Am I consciously promoting positive experiences for horses in my training sessions? What indicators do I use to assess positive emotions in horses?
5a) What specific behavioural or physiological indicators do I use to notice when the horse is shifting from a manageable state of arousal into anxiety? How quickly do I adjust my approach in response?
5b) How confident am I in distinguishing between signs of alert engagement and early stress signals? How does this influence the speed, proximity, and intensity of the desensitisation process?
6a) Do I apply pressure in a way that is immediately released upon the desired response? How do I track my timing?
6b) What strategies do I use to minimise reinforcement delays? How do my delays affect the integrity of my training?

6c) In what ways do my implicit habits override what I know about learning theory (patting instead of scratches which the horse values)?

7a) How intentionally do I condition the horse to respond to my seat as a precursor cue before applying lighter aids? Am I consistent enough for the horse to form reliable associations between these signals?

7b) How do I monitor for unintended associations (overshadowing), and what steps do I take to avoid them?

8a) Do I systematically deconstruct complex behaviours into small, achievable steps with clear reinforcement at each stage?

8b) Do I shape down as often as I shape up? Could I shape up and/or shape down more quickly? Improving consolidation of learning.

9a) Are all my aids clear? Discreet? Biomechanically timed (start of swing phase) for the horse's understanding and physical coordination?

9b) How do I assess and refine aid clarity to prevent ambiguity and overlapping aids?

10a) Am I fostering self-carriage by avoiding coercion and minimising continuous pressure or signalling?

10b) What criteria do I use to evaluate whether the horse maintains speed, line, and outline autonomously?

LEADING CULTURAL CHANGE
COMMUNITIES AND INFLUENCE

Every equestrian belongs to a community: a yard, club, discipline, breed society, or online groups. Our communities shape how equestrians talk about horses, interpret behaviour, and judge training choices. Communities are powerful ecosystems of beliefs, behaviours, and identities. Choosing to enhance horse welfare outcomes is choosing to evolve equestrian culture by learning to change the stories communities share. Humans are social learners, and group consensus gives people a sense of safety. Change—even ethical, evidence-based change—can feel like a threat to belonging. That's why the role of visible practitioners, trusted educators, and informal leaders is so crucial to positive change for horses. We act as bridges between what is familiar and what is possible. When someone in a community takes the first step—loosening a noseband, releasing rein pressure for a desired response, questioning outdated methods—it invites curiosity. When that same person explains their reasoning, shows results, or maintains humility, it invites trust. This is how scepticism transforms into stewardship: one conversation, one demonstration, one role model at a time. To create welfare-first community narratives, we must

- Normalise change as growth, not failure
- Celebrate curiosity over certainty
- Reward welfare action, not just performance outcomes
- Make evidence-based practice visible, repeatable, and inclusive

It is also essential to understand resistance as part of the process. Resistance is rarely about the facts—it's about identity, fear of being judged, or discomfort with ambiguity. Compassionate responses

DOI: 10.1201/9781003608066-4

to resistance can keep the conversation open. Rather than saying, 'You're doing it wrong,' ask 'What would it take for this to make sense to you?' Welfare Horse Sports thrives when communities become places of exploration, not just tradition. When asking questions is more valued than having all the answers, stewardship replaces status, and leaving horses better becomes the shared mission—not just an individual act. That's how movements grow—not through mandates, but through meaningful modelling. And it starts wherever you are.

LEAVE HORSES BETTER CoP

Between 2017 and 2023, I ran an online Community of Practice (CoP) centred around leaving horses better because of our interactions. In 2025, former members were independently interviewed, providing them with a voice to share their lived experience of being in the community and *doing* the work to leave horses better. Reflections were captured, thematically analysed, and evaluated and presented at the 20th ISES conference, (2025). Three key areas of being in a CoP were identified:

1. Challenging traditional norms through peer-led learning
2. Applying First Principles of Horse Training (ISES) in daily horse training
3. Use the Five Domains to redefine success beyond competition metrics

Preliminary findings indicate that engagement in a CoP accelerates the application of First Principles (ISES) and reshapes professional perspectives on horse welfare. Members reported:

- Increased confidence in using evidence-based horse training knowledge, including reinforcement timing, removal, and additional reinforcement
- Greater willingness to challenge outdated methods, particularly within riding schools and coaching programmes
- Stronger professional engagement in welfare advocacy, practising transparent communication with students, clients, and the broader equestrian community

This research found a CoP to be effective in driving cultural change, equipping professionals with the skills, confidence, and community support to apply evidence-based knowledge and advocate for improved welfare policies. CoPs offer a collaborative solution for integrating equitation science into equestrian education.

> You're always practising something. So you're either practising upholding the world as it is, or you're practising shifting into the world as you want it to be.
>
> Adrienne Maree Brown, as cited in Triple WellBeing® Framework (n.d.)

OBSERVABLE BEHAVIOURS: INDICES

The following are observed when applying First Principles (ISES):

- **Enhanced Horse Welfare**: Prioritising the horse's nature (provide social partners by removing bars between stables to intentionally provide the positive experience of a deep relationship between two horses) and mental and sensory abilities, and combining procedural knowledge with *critical anthropomorphism*, a concept introduced by Gordon Burghardt in the mid-1980s, emphasises the importance of intersectionality thinking. It is an approach 'using our human experience to recognise animal suffering by checking our immediate intuitions about an animal's subjective life against which we can learn from more objective scientific studies' (Morten et al., 1990). Think of critical anthropomorphising as 'guard-rails' for navigating inevitable anthropomorphic tendencies. In the ongoing 'Future Horse Project' from the University of Melbourne, Australia, researchers explain anthropomorphism is prevalent in society, often providing equestrians with a sense of personal identity, a way for them to make sense of the world. Anthropomorphising may motivate a connection with horses and influence the development of a sense of attachment and social positioning of the horse, for example, 'fur child.' It also may influence cultural perspectives such as 'moody mares,' brave cross-country horses, and horse-racing heroes. Anthropomorphising may also lead to equestrians justifying outdated beliefs and therefore behaviours misaligned to their mental abilities (First Principle # 3), for example, the naughty horse. Applying First Principles helps equestrians identify a once-held belief (a horse is lazy/naughty/needs to respect you). When we approach horse behaviour with a critical anthropomorphism lens, we see the suffering for the horse and check in with other possible realities (by applying scientific knowledge such as the Ridden Horse Pain Ethogram [RhPE]) for the horse; pain, fear, and confusion. Scientific knowledge of a species, *coupled* with our ability to think critically, is essential. Critical thinking is being objective, systematic, and doing a rational analysis or evaluation of factual evidence, informing a judgement. Whilst anthropomorphism is prevalent and uncritical, being aware that we are thinking anthropomorphically and recognising this is a first step towards enhancing horse welfare. The next step is developing the ability to think critically. Thinking critically is an observable behaviour. Take down the horse's history, write down the specific behaviours (drop cultural short cuts i.e., lazy, 'spooky,' naughty, moody) and replace them with what you actually see. For example, a horse 'spooking' is one who has slowed (stopped abruptly) and then rapidly turns his forelegs (away from the aversive stimulus). Combine your knowledge (written down) with *species-specific* knowledge, recognising that the most prevalent strategy for horses (species-specific behaviour) is avoidance. A fearful horse is highly motivated (emotionally) to avoid any stimulus perceived by the horse to be a potential threat to their survival. You know, the bucket that has now arrived at the side of the arena is just a bucket. Your horse even eats out of the exact same type of bucket (shape and colour). Yet today he slowed at the sight of the bucket, head and neck tense and raised, ears locked forwards in the direction of the bucket. In a nanosecond, you are both facing the opposite direction to the bucket (slowed and turned). The high levels of arousal escalated from the first visual picture of the bucket because it was not there yesterday. Horses capture our environments as photographs. Horses are context-specific because it is necessary in evolutionary terms that they notice changes in the environment. When a bush rustled and a predator jumped out millennia ago, responding to any bush rustling by activating fast legs and creating distance was how your horse survived frequent enough threats of predation. Knowing species-specific behaviour (Principles 1–10) *with* critical thinking (collect and evaluate the behaviour through a species-specific lens) is how equestrians form a *cautious inference*. An approach that places empathy for the horse *alongside* evidence-based knowledge, enhances the understanding of the experience of horses, whilst avoiding the pitfalls of anthropomorphism.

- **Improved Training Outcomes**: The clarity of communication between horse and rider, through the correct application of learning theory, results in horses being able to flourish (as defined by the positive animal welfare [PAW] definition). By providing mostly predictable (light aid/cue/pressure) and controllable (rider removes pressure for the correct response) cues in *all* interactions with humans, equestrians create competent horses by applying the shaping Principle (# 8) and welfare horse sports mindsets that train to set horses up to succeed. This is observed by stepping the horse back so the hindlimb the trainer is targeting to adduct (leg-yield) is at the start of the swing phase, that is, the hindleg is back behind the other hindleg (not under the spine or equally placed next to the other hindleg), or when you are training your horse to stand at the mounting block, set them up to succeed by bringing in the environment. I start mounting training along the wall of an arena. This helps the horse answer my posture cue (raised hand), followed by the training device (long whip) reaching over to the back of the saddle (cantle) and fast tapping the cantle, followed in the next second by tapping the top of the right hindleg. The horse controls the light fast taps by stepping away from the wall of the arena. Setting a horse up to succeed looks like helping horses experience the correct response (whatever it is we are shaping towards training). In this example, I am shaping the visual cue of my raised hand, which is the predictor, that light taps of the saddle follow (aversive sound), followed by light taps experienced on the top of the right hindleg. What the horse is experiencing is *predictable* (light taps), followed by *controllable* (release of taps for stepping away from the wall)—Principle 6, the correct use of operant conditioning via removal reinforcement. I am a combined reinforcement trainer, so you will see me bridge the behaviour from the horse that I want more of (stepping away from the wall) via a verbal bridge. Mine is a 'cluck' followed by a scratch on their 'sweet spot' (where they find touch/scratches optimal) or the delivery of a fibre nut. With repetition, normally on the third rep, my hand raised predicts for the horse the sequence: cantle taps preceding top of hindleg taps. Due to the mode of learning; classical conditioning (Principle 7), now the horse associates raising my hand with the sound, then the feel of the right hindleg being touched, moving away from the wall/side of the arena. The observable behaviour of applying First Principles in training a horse to stand at the mounting block requires knowledge of the First Principles *and* recognising in your own horse training, clear communication from you, in order to provide improved (effective) training outcomes. Making mistakes, both horse and human, is normal and essential to expanding your own and your horse's resiliency. You will observe yours and others' mistakes during the application of the First Principles. This is part of the process of learning—normal. We need to experience mistakes (speed of taps, lightness of taps, and timing to stop tapping) to provide us with feedback on our *operant* contingency, that is, light aid, motivational pressure to create desired behaviour, and the release/off of the pressure for the desired behaviour. Operant contingencies are observable and indicate/demonstrate your own level of skills and stage of learning. The horse's response to your mistakes also provides observable behaviour of the application of the First Principles. As a result of our mistakes, the horse experiences something discovered in the field of neuroscience—'prediction errors.' A prediction error is a mismatch between what we expect to happen and what actually happens. As an observable behaviour of First Principles (ISES), prediction errors offer equestrians feedback as to whether our communication/training is clear. Clarity of communication is observable, as we witness how horses receive our cues/communication. Let me explain. When you anticipate your horse will respond to a cue, and your horse reacts differently, this is an observable prediction error. Prediction errors explain how your horse (and humans) adjust behaviours because of/based on our experiences. Prediction errors are fundamental in learning theory, neuroscience, and psychology because prediction errors update our *future* expectations. When a trainer cues a horse to lead, the hand goes forwards towards the chin, the cue to accelerate for the horse. The horse walks forwards. This is an example of an expected outcome, therefore no prediction error. When there is a different outcome from what is expected, the horse walks backwards, and a prediction error occurs. Horses (and humans) use prediction errors to guide learning and inform our future expectations. When an outcome for the horse is perceived by the horse as *better* than expected, the behaviour is reinforced, and scientists call this a *positive* prediction error. When an outcome is *worse* than expected, horses (and humans) adjust what we observe (behaviour) to avoid the same outcome in the future; called a *negative* prediction error. Horses use prediction errors to update learning and future expectations. Applying the First Principles (ISES) increases the observables of positive prediction errors, as horses experience less than expected pressures, faster release of pressure, and if you apply combined reinforcement (removal of pressure followed by the addition of a scratch or nut) you facilitate improved training outcomes, because your horse's experience is better than expected training outcomes, and research reports the animal, on experiencing a better than expected outcome, positive prediction error, gets a dopamine 'lollipop.' Observing the correct application of First Principles also looks like releasing your tacked up horse (at liberty) in the training arena (place you train) and observing what your horse does next. I have clients who reinforce the mounting block. What you observe could be interpreted as agency (horse chooses) to be ridden or not, or applying the First Principles you are able to identify the context in the arena where secondary positive reinforcement is delivered; at the mounting block. Apply a verbal or physical bridge such as a mouth 'cluck' or clicker followed by food (application of operant conditioning, specifically additional reinforcement) at the mounting block and for standing whilst the rider mounts, an observable behaviour of applying First Principles 6, 7, and 8; operant and classical (conditioning and shaping). When your horse takes herself to the mounting block, she gets reinforced for that behaviour (some trainers escalate the attractiveness of the reinforcer, food, by providing a flavoured lick-it tub) this is then immediately followed by the rider mounting. The lick-it acts like a 'station'—a context where something happens to the horse. In this example, the rider mounts. With repetition, now the mounting block is the context for the reinforcement of the observable behaviour of standing immobile at that mounting block, context specific. To onlookers, the behaviour observed is a horse choosing to be mounted. Applying First Principles has trained the horse at liberty while approaching, standing, and rider mounting in the context of the arena and only at that mounting block *for* reinforcement. When does mounting training at liberty become a horse experiencing choice? On the days your horse does not do the observable behaviour previously reinforced (trained) and instead stands by the gate or wanders around the arena, you take your horse out. What is observed is the horse choosing not to be reinforced by you, that day. Perhaps the environment is more reinforcing, for example, a new horse arrived on the yard who is in the field opposite the arena, or perhaps the horse chooses not to be reinforced due to some painful lived experience. Because they are not carrying out the learnt previous observable behaviour, you can now respond to what you've witnessed and investigate, be your horse's detective, and monitor her animal-based indicators in the Four Domains to make a cautions inference on the change in observable behaviour on entry to the arena. Responding to how the horse *feels*, every single day, is how horses experience our ethicality (responsiveness to the horses lived experience) and where possible, offer better than expected (dopamine lollipops) or as expected predictions that help your horse navigate their world, because it is both predicable and controllable. Some introspection of what you have just observed is useful at this juncture. When you train applying First Principles specifically combined with reinforcement, you can offer a sense of agency (choice) to your horse. Think about the days (how often and when in a week and season)—do you observe your horse choosing not to be reinforced for the previously learnt behaviour? Exploring the answers to these critically reflexive questions help equestrians naturally bring into focus evidence based knowledge and help individual horses have their species specific needs met (telos).

- **Increased Safety**: You can observe how your horse feels through recognition of their language signs with calming signals to assess if your horse feels safe. A relaxed, alert, and optimal state for the process of learning can be observed, *and* the application of First Principles optimises the safety of the rider/handler. I liken the application of First Principles (2018) to a safety checklist, such as a pre-flight check or Ministry of Transport (MOT) certificate (former UK government department responsible for transportation). An assessment of vehicle safety, roadworthiness, and exhaust emissions is required *before* it is legal to drive a vehicle. We do not drive a vehicle legally whose brakes only sometimes work or steers only to the left! First Principles (ISES) provides you with an in-hand and under-saddle unwritten check-in that your horse is optimal to learn (regard for emotional state, Principle 4) and therefore train. Before, during, and after training, we can also observe a horse's (and human's) arousal levels. By training specific behaviours using removal and additional reinforcement correctly, a desired behaviour such as immobility ('park' response) when previously learned in low arousal environments offers behavioural change/modification, resulting in arousal levels changing from high to lower/low. For a flight species, such as the horse, training to stand still is the equivalent of 'switching off their flight switch.' When we look into our past as horse trainers, cowboys knew this by inventing the 'stock' and 'hobbles.' In today's horse-centred landscape, a welfare horse sports mindset chooses First Principles (2018) to *train* immobility (no coercion). I am often asked the question: what can be done when observing your horse escalate from feeling safe to all the behavioural indicators of now feeling fearful? Perhaps the question we should be asking is: Did I recognise the stimulus for escalation? If so, why did I watch/observe? The answer will require introspection and critical reflexive thinking. When we have evidence-based knowledge to change how horses feel towards aversive stimuli (and previously trained the learned behaviour in a safe context), why don't we think to apply First

Principles (2018) and help horses stay below their individual anxious 'threshold'? Knowing there are baroreceptors in the head of the horse (immediately below the poll, as in all animal species with long necks, e.g., giraffes) when the poll lowers to the same height as the heart (left side of the chest), the horse's blood pressure slows, in turn, slowing the heart rate. Knowing better (how to change how your horse feels, anxious to relaxed in response to aversive stimuli) and doing better for your horse is observed through increased safety, helping horses feel safe, as evidenced by their mental and physical relaxation. If you are unable to get a clear response to accelerate, decelerate, turn, and yield forelegs and hindlegs from light and immediate signals in-hand, do not proceed to under-saddle. This is your safety check-in. If your horse is heavy and/or delayed in-hand, they will be heavy and delayed under-saddle. Applying the shaping Principle (# 8) means training first in-hand, then as the behaviour you observe shapes up (improves via witnessing lightness and responsivity, first one step, then two steps/strides in canter, then multiple steps), you can progress to under-saddle. Remember to start again at the beginning, that is, basic learned responses: go, stop, turn, and yield (accelerate, decelerate, abduct, and adduct). Horses are context-specific (Principle 2), so embrace the crucial skill of shaping *down*. This is when you decide, as a horse trainer, to go down (some people prefer 'back,' but I have learned the word 'back' is a reactant for some equestrians, as it reminds them of their positionality and experience, the word 'back' as a framing of failure). Identifying the word 'back' as a human construct for the opposite of progress is healthy. Pause at this point. Think about the construct for you of the words 'shaping back' versus 'shaping down.' Which do you prefer? And why? Remember your horse is just standing in the here and now, possibly waiting for his next opportunity of reinforcement. Or perhaps her affect (emotional state) has changed, helping you decide that the next step in your shaping plan is to shape down, offering another opportunity for your horse to learn *you* are the source for providing a better than predicted outcome and another dopamine lollipop! Shaping is all about finding the smallest unit of behaviour to reinforce. Stay agile in shaping because you attach no value judgement to shaping down or up, both are necessary to shape/train the new behaviour you desire. Shaping down expands confidence. Confidence building is the foundation for resiliency and expanding your own and your horse's resilience. Observing shaping up and down depending on the competence and confidence of the horse, is evidence a horse trainer is paying attention and capable of being responsive to the mental and physical state of the horse. Shaping down *and* up is how we help horses feel safe and how we stay safe. Always shape down because we respond to Principle # 4, the emotional state of the horse, not just the behaviour observed.

RESILIENCY BUILDING

Applying First Principles (ISES) sets the horse up to succeed. What we then observe as an outcome of competence is horses curious to try a new response in order to experience the release of light pressure and receive the bridge (cluck) and a scratch/nut. During challenging times such as medical interventions (injections, wormers, or oral meds, etc.) the observable behaviour is when your horse endures the aversive procedure (does not freeze, fight, or flight), for the consequence of the additional reinforcement. Learning how to set horses up to succeed can be experienced with the support of peers inside a CoP.

LHB: LEAVE HORSES BETTER

LEAVE: *Lean* **into curiosity.** Ask 'What am I missing?' Get curious about a horse's lived experience *during* your interactions. Place your focus on what they are experiencing (use the Five Domains 2020 framework). Instead of asking, 'What is it like to be a horse?'—ask, 'What is it like for a horse to be a horse?'

HORSES: *Help* **horses resolve negative experiences.** Do pragmatic innovation. Decide to *do* small realistic actions with impacts for horses by helping them resolve negative experiences related to their specific species behaviour (friends, forage, freedom, *and fortitude*). Fortitude, the fourth 'F,' was founded by Dr Andrew McLean in 2024 during podcast episode 5 of *The Other*

End of the Reins with Dr Andrew McLean and Lisa Ashton. Listen to the episode and recognise that fortitude is a practice that gets expanded (like resilience) because of adversity. The sun is harmful during the hours of 12–2 pm, and we resolve this negative experience by going in the shade, or if we don't have that level of agency, we mitigate the harm from UVA exposure by applying sunscreen. *Helping* horses requires more than knowledge; it requires *procedural knowledge*. We must bring what we know into action in order to help horses resolve their negative experiences. Practice observing your horse's whole body. Arousal plus valence (positive or negative experiences) can signpost us towards making a reliable inference about a horse's emotional state. If a horse is avoiding stimuli (including people), recognise avoidance behaviour as an indicator that the stimuli is a negative experience for that horse. If you are struggling to catch a horse, this is your indicator of a negative experience. What comes after catching does not provide a positive experience. If it did, your horse would approach you, the first in a chain of consequences for being caught and the following positive opportunities. Negative experiences are essential to living a good life. It is functional to an individual's motivation to resolve the negative experience. Hunger, pain, and isolation must be experienced *and* resolved in order to experience a good life. The welfare horse sports mindset focuses on noticing negative experiences in order to help horses resolve them, thereby expanding resilience. The opposite of a positive mental state is a horse presenting apathetic to attractive stimuli, avoidant, or, when unable to resolve negative experiences, confusion (conflict behaviour) and/or frustration. Work through each domain and apply First Principles (ISES). Helping horses resolve their (not our) negative experiences also helps us. Researchers experimented with an online microcharitable giving intervention, where voluntary donations of one Chinese cent (¥0.01 or $0.0014) over a 2-month randomised controlled trial with depressed individuals showed that the intervention group exhibited significantly greater improvements in both depressive symptoms and emotional positivity, and that emotional positivity mediated the intervention's effect on the reduction of depressive symptoms. The impact of these low-cost findings provides potential for an accessible intervention for the prevention of human depression.

REFLEXIVE PRAXIS

Can your horse explore their own curiosity? That is, self-regulate, through the species' need to chew (access to forage throughout the day/evening), and receive/provide mutual grooming?

> I grow little of the food I eat, and of the little I do grow, I did not breed or perfect the seeds. I do not make any of my own clothing. I speak a language I did not invent or refine. I did not discover the mathematics I use. I am protected by freedoms and laws I did not conceive of, legislate, or enforce. I do not build the devices I use. I did not invent, patent, or perfect the technologies I use.
>
> I love and admire my species, living and dead, and am totally dependent on them for my life and well-being. Sent from my iPad.
>
> Jobs (2023)

BETTER: *Bring* **opportunities for positive experiences.** The first word of our horse-centred mindset is intentionally welfare. Observe the word 'welfare' as two smaller words: 'well' and 'fare,' 'well' meaning feeling healthy (physical state) and happy (mental state). 'Fare' comes from an old word that means travelling or going through life. So, when we say 'welfare,' it really means 'fairing well' or 'living in a good way.' For a horse, that means having friends, access to food (forage), and space to move freely. Leaving horses *better* stretches 'welfare,' horses fairing well, (inferred from animal-based and situational-based indicators of mental and physical state of the horse) to *well-being*. Horses are provided opportunities to experience positive experiences. The Quality of Life framework explores animals living a 'good life.' Leaving horses better is at the centre of well-being, a sense of how can a horse *be*? What does it mean *as a horse* to thrive in the way that horses do? When we allow horses to *be* as healthy (mentally and physically) as they can *be*, as in nature, horses' competence and flourishing are in abundance, thriving. The Well-Being of Future Generations Act (2025) in Wales is a recent piece of legislation established to create a Wales that current and future generations want to live in. The Well-being Economy Alliance has been working for years across the world to redesign economic structures to serve people and planet, delivering quality of life and flourishing for all, in harmony with the environment, not in combat. Knowing humans are working creatively and innovatively, we can see how we can get ourselves unstuck from the equestrian 'this is how we've always done it' mindset. The Well-being Economy Alliance has been working on undesigning and redesigning for communities to thrive in the world; we have global examples of schools, leaders within existing models of education, changing the culture. Observing different ways of schooling is changing the education culture by putting healthy child development at the centre of education and bringing a play-based curriculum into schools because it makes sense for healthy child development, with parents and communities saying yes! Where a well-being curriculum does not tick an Ofsted inspection box, educational leaders explain to the inspectors that their school is *prioritising* well-being. These schools recognise they are not beholden to a system failing children for the rest of the school's existence. They recognised their capacity to change. Returning sport horses to well-being, thriving as a horse, is the impact of a welfare horse sports mindset. Returning horses to be able to thrive is the impact of leaving horses better. The 2019 Pulitzer Prize winner for fiction was the novel *The Overstory* by Richard Powers, focusing on environmental themes and the weaving together of scientific facts about trees with human drama. Leave horses better is the overstory of our welfare horse sports mindset. Lean into curiosity (addressing equestrian culture 'this is how we have always done it'), help horses resolve negative experiences (end suffering of horses used in sports by seeing the systemic brutalisation because we are weaving together scientific frameworks that centre species-specific knowledge [Five Domains [2020] and First Principles [ISES] via procedural learning) to then show up and bring opportunities for positive experiences, nourish our horses and keep ourselves connected to their being well. That is the place that keeps us, equestrians, well. *The Overstory* asks readers to reconsider their relationship with nature and their role in preserving the world's forests. The human stories in the novel form the 'understory' whilst the overarching narrative of trees and forests represents the 'overstory.' Combining evidence-based knowledge with a profound human commitment to enhancing the lives of horses—leave horses better—is our welfare horse sports mindset 'overstory.' The environment (planet), equines (horses), and equestrians (humans) cannot flourish or, in context to positive horse welfare, *be* well (well-being) if we keep telling ourselves the use of horses in sport status quo story. Joanna Macy's *Three Stories of Our Times* explain our horse sports past, present, and potential for the future. The story 'business as usual' is the story where we carry on as we are, blind to immense societal attitude change, believing horse sports will all be fine. The second story is 'the great unravelling,' the story that we are all doomed if horse sports reform horse welfare, claiming 'next we won't be allowed to ride horses.' The final story we equestrians could engage with through our welfare horse sports mindset is 'the great turning.' This is the story in which we can all make a difference to create a healthier (mentally and physically) world for horses, humans, and the planet. Equestrians keeping on living the 'business as usual' story is understandable because it's the story that is the most comfortable and perhaps the least terrifying. Anything positively disrupting the status quo, or opening equestrian eyes to what is going on for horses (evidence-based knowledge), is scary. This is a human condition, not just an equestrian response to positive change. Because the great turning story is inconvenient, it is therefore understandable why people feel uncertain about welfare reform and reject a new story, that is, 'the great turning.' When we choose to open our eyes about the current state of our planet, we are seeing more frequent extreme weather events. Climate change is not something we can close our eyes to. If we stop and look at it, the business-as-usual story does not work. If we then go into the story of 'the great unravelling,' it is totally understandable that young people experience a psychological response to real and perceived environmental threats known as 'eco-anxiety.' Providing understanding to 'the great unravelling story,' sharing that any welfare reform will result in the end of riding horses, comes from a place of fear and distress, driving extreme responses, blocking bringing people along. When we turn to the third story, 'the great turning,' bringing people with us, showing a better way for using horses in sport/grassroots (positive horse welfare), we do leave horses better because of our interaction. Knowing equine well-being is not optimal in horse sports, the great turning story is our collective story of our welfare horse sports mindset.

The overstory to *leave horses better* has its understory in intentionality and the actions we take for our horses or, at the very least, think about the impact of our actions today on our great-grandchildren's great-grandchildren interacting with horses. This seven-generation thinking is an Indigenous principle that our actions impact everything, everyone, and everyone that's coming, including horses. Leaving horses better is thinking beyond ourselves. An invitation to see that life is much bigger than us, helping horses *be* well, not just thinking about ourselves, but recognising our equestrian legacy is also our ancestry and is part of our equestrian story. It is about honouring the horses who came before the horses we leave better today. The horses who have paved the way for a world turning towards doing welfare reform. Whilst leaving horses better through our own interactions, equestrians are paving the way for future generations to leave horses better, creating an equestrian world better than the one today. Equine well-being and the impact we have on future generations of equestrians enhances our well-being. The Future Generations Act (Wales, 2025) is thinking longerterm. The absence of evidence-based knowledge in horse training methods means nothing can change (welfare reform). When the horse training systems educating equestrians are held up by an outdated education curriculum (absent of the *application* of learning theory/First Principles [ISES]), we prevent generational progress and the change that leaves every horse better because of our interactions.

THE IMPORTANCE OF WELL-BEING

A country situated between China and India, Bhutan, measures its wealth by an index called Gross National *Happiness* (GNH). In 1629, ancestors created a code: 'if the government cannot create happiness for its people, then there is no purpose for government to exist.' The code informs Bhutanese law, working towards happiness for all sentient beings and cultivating a culture derived from the four pillars of GNH and nine domains supported by their structural systems. In contrast, UK wealth is based on Gross Domestic Product (GDP), an economic measure of wealth. Structural systems that enable everybody to thrive are at the heart of a country's framework, an appreciation that everybody matters to that country. The index is personally mapped, so any indicators of ill health in the nine domains are given priority. Leaving horses better is founded on the Five Domains and First Principles (ISES), shining a mirror on the web of well-being. Planet, animals, and people are interconnected; when one thrives (horses), the 'new happy' framework by Stephanie Harrison explains the attention to service to others (horses), helping others rather than helping ourselves (which is one of the greatest sources of unhappiness) by contributing to another's well-being—the horse—through pragmatic innovations to leave your horse better. Little actions add up to a world where more and more horses get to experience *being well*. According to Harrison, helping others is something we have to devote ourselves to. The sense of loneliness, despair, and helplessness experienced by people today is a result of thinking we don't owe anything to anybody. In reality, we owe a lot to everybody, not least our horses. The interconnectedness of equines, the environment, and the equestrian community provides a sense of purpose, knowing our lives are making a difference to the lived experience of horses, people, and the planet.

 REFLEXIVE PRAXIS

Watch *The Dancing Man* video here: *https://www.youtube.com/watch?v=fW8amMCVAJQ*

A man is dancing on the side of a hill at a festival. Everyone else is sitting there looking at him. Then, after a while, people start to join in with him. When enough people stand up to *do* welfare reform for horses (start with having honest conversations about horse training needs to change), it becomes okay to join. In so many aspects of society right now, we are experiencing shift points. More and more equestrians are saying in their communities that this does not work for horses, equestrians, and the environment any more.

This is how positive tipping points happen, with changed beliefs cascading into communities and changing future human behaviour.

In 2025, the FEI invited Dr Andrew McLean to collaborate on the FEI Equine Welfare Strategy; 'My remit is to foster the uptake of the ISES First Principles of Training to promote optimal welfare and sustainability in horse sports. Equitation science offers unique potential for achieving the goals of horse sports, yet within the framework of a good life for horses. It is now more important than ever that riders, trainers, coaches, and judges need to know what they don't know.' Dr McLean is the 'dancing man' and has had many 'first and second' followers around the world for two decades, upstanding for horses by doing self-leadership, advocating for horses by serving horses to experience positive welfare through the application of evidence-based equitation (EBE) (applying the First Principles, ISES). McLean is at the cutting edge of welfare reform, asking all equestrians

'If you come to know better, would you do better?'

 REFLEXIVE PRAXIS

Take time and consider this question. Journal what comes up for you.

Self-deception in horse training reflects the same psychological processes seen in magic, misinformation, and identity protection. Neuroscientists call this *metacognitive impenetrability*—our brains create explanations for behaviour without our awareness, leaving us convinced we 'see' or 'know' something that may not be real. Magicians exploit this through illusions like the *bluff vanish*, where audiences confidently report seeing objects that never existed; equestrians do something similar when they misinterpret a horse's learned behaviour as 'naughty' or 'disrespect' rather than a functional response to seek avoidance (maybe due to pain, fear, or confusion).

Researchers have discovered metacognitive impenetrability, the actions taken by our brains of which we are unaware. It springs from our motivation to make some kind of sense of what we are experiencing. So, making sense of the unwanted behaviour, a 'barging' or 'pushy' horse is now being perceived by the owner as disrespectful. Believing any horse is disrespectful makes equestrians unreliable guardians. Neuroscience researchers investigated techniques that trick the brain, going into the scientific silo of neuromagic: applying magic (tricks) as a way to make sense of perception and cognition. There are even organisations that put scientists and magicians together—the Science of Magic Association (SOMA). The idea behind the organisation is to bring together performers in the magic performance community with researchers working on collaboration, in order to research different kinds of illusory methods to explore different ideas about human cognition, and then, optimally, the performers can also take away something to design new tricks and expand their repertoire, with different ways of thinking about how they present and perform magic. This is known as 'magic thinking.' What do magic tricks have to do with welfare? In one magic trick called 'the bluff vanish,' magicians land into our retention of vision ability and specifically our retention of vanishes, for example, you hold an object and pretend to take it from one hand to another with a bluff vanish, as if there isn't an object. The trick is a pantomime. Scientists use the foundations of this magic trick to collect a distribution of responses. The study wanted to find out if people would give testimony, like witness reports, of objects that never existed. By using pantomime, researchers pretended to take something out of a cup. In the critical trials, there was no object, just pantomime. They were making nothing disappear. The findings showed 30% of people confidently swore there was something that the magician had made disappear; 10% of those provided specific visual details: a silver coin, a red ball held up to the camera that caught the light—yet there was nothing there. It was unclear from this study if people had a genuine perceptual experience in the moment, like a mini hallucination, or if they had a false memory of the object. Researchers in this field call this a naturalistic, dynamic scene. The scientists went on to perform some live bluff magic tricks, showing humans give false testimonies, unclear if people experienced a false perception or a false memory in just seconds. These findings help to navigate the endemic that is misinformation, giving false verbal feedback within seconds, not days. Psychologists who study human thinking and perception have looked back to the 1800s, where people integrated magic methods with behavioural research methods. In 1848, the Fox sisters is a story of children hearing noises in their house, knocking noises. People came and couldn't find the sources of the noises. The noises kept happening. Eventually, the children worked out a communication system with the noises, and they started talking to it. The Fox sisters, Margaret and Kate, became famous and made a lot of money. Their older sister, Leah, became their business manager, and they held public events that drew in crowds. Over time, Leah became estranged from Margaret and Kate, and Margaret and Kate developed alcoholism problems, along with financial disagreements, leading Leah to hold a public event at the New York Academy of Music to expose their hopes. She demonstrated that they used *toe cracking*. They cracked their toes to create the sounds. She did this in a theatre where it echoed around the room and made the sounds people recognised from the demonstrations. What is really important for horse welfare is that believers held on tighter to their belief in the paranormal, despite one of the sisters taking accountability for the deception. When equestrians *unconsciously* convince themselves of horse training traditions, despite the sizeable quality of evidence-based knowledge (meta-studies sharing what we now know and evidence for optimal mental states of horses in sport at this time), *self-deception* is hiding in plain sight. Unlike a magic show, people are aware the entertainment revolves around humans experiencing deception, some indulging in self-deception, equestrians deny, rationalising away the significance of opposing peer reviewed evidence and logical arguments. Doubling down on traditional beliefs, especially in the face of evidence-based knowledge, is the same as onlookers at the Fox sisters hoax reveal refusing to accept the noises were the sisters toe cracking. This example of the human condition to self-deceive is the same condition equestrians experience when attached to traditional horse training practices that are sub optimal or harm to horses' mental and physical wellbeing. Denial of evidence, demanding more evidence, and ignoring and dismissing robust evidence, because it contradicts with one's beliefs or winning desires is self-deception. Justifying, explaining away the inconsistencies between evidence-based horse training (First Principles, ISES) and traditional equitation beliefs, according to self-deception researchers, can be without conscious awareness, making it even more challenging to recognise, especially when equestrians are making money living the deceptions. That is how magicians are able to make a living as stage performers. According to researchers, magical thinking is most likely a relic of our evolutionary past, the early brain struggling to detect and then impose order on a chaotic world, where seeing agency and rustling leaves could mean the difference between life and death. A false positive is a trivial error, but a false negative could be fatal. Self-deception is thought to be an ancient impulse encoded in human cognition. The human brain is flawed. It is evolved for survival, not for truth. That is why we invented the scientific method to add a few more steps to our thinking *before* reaching conclusions, writing laws, and creating reform. All manner of deceptions exist as part of the human condition. Even when we believe we are immune. Especially when we believe we are immune. The history of psychology has shown when controlled experiments are conducted, when double blinding, peer review, and rigorous testing is applied, the illusions created by magical thinking collapse (Clever Hans phenonomen). They vanish. Self-deception is a cognitive condition that enables fraud, for riders to be judged at 70 percent and above, and even win medals. Instead of humans grieving at being exploited at the fraud, in deception and self-deception by equestrians, horses pay the mental and physical price. As psychologists joined magicians in their systematic debunking, equitation scientists join horse trainers and discovered that what we once believed (did the best we could with what we knew at the time) as foundational to our equitation practices, turns out, is in fact, less than optimal for the well-being of the horse. This is not usually malicious but an unconscious act of identity protection: dismissing evidence preserves belonging (to equestrian identity), esteem, and continuity with tradition, even at the expense of horse welfare. Psychologists remind us that the human brain evolved for survival, not truth, which is why magical thinking and self-deception are so persistent. In magic, the cost of deception is applause; in equitation, the cost is borne by the horse's mental and physical well-being. The role of equitation science, like the scientific study of magic, is to expose illusion with method, making visible what self-deception hides, and ensuring that evidence—not tradition—guides practice. Equestrians' intentions are good, their hearts big, compassion great. There is no desire or intent to harm horses, but a form of self-deception.

WELFARE REFORM

As we have journeyed through the welfare horse sports mindset, we have centred reflexive praxis with the practice of critical thinking, a continual process for critically examining our beliefs, actions, and impact, with the aim of developing understanding through growth, in order to enhance horses' lives. Reflexive praxis goes beyond reflecting on our experiences. It is understanding our identity, biases, fallacies, and yes, even our propensity for magical thinking (self-deception) as we have explored the mindset of a welfare horse sports equestrian and what observable behaviours of a welfare horse sports mindset look like.

REFLEXIVE PRAXIS

1. *Critical Self Detective*: Frequently assess and question your beliefs, values, and actions both around and not around horses, gaining a deep dive into your self-awareness.
2. *Contextual Awareness*: How social equestrian culture (heritage and contemporary) influences your horse training practices and your horse–human interactions every day.
3. *Adaptive Learning*: Applying insights gained from reflections, creating catalysts to adapt and improve *for horses* and equitation practices.

Before we take a zoomed-in perspective on welfare reform for horse sports, let us start with 'zooming out.'

We have explored curiosity and learned that being curious is the start of positive change for horses. Combining horizontal curiosity (zoomed out) with vertical curiosity (zoomed in) embraces a diversity of thinking to solve complex challenges. When we are curious, we have an appetite for learning and take steps *outside* our equestrian beliefs. Let us now take a look outside of equestrianism to explore what we can learn from the field of medicine.

Applying a zoomed-out curiosity, I want to take a shallow dive into evidence-based medicine (EBM). How did the EBM movement overcome initial resistance among medical professionals? Then, a 'zooming in' on welfare reform for horse sports, both inside and outside of sport, and within our grassroots centres, which serve as the first touchpoint for future equestrians and a welfare horse sports mindset.

Before EBM in the early 1990s, medical decisions made by practising doctors came from clinical experience. Like horse trainers, before EBE emerged in 2015, led by the ISES, doctors made decisions for patients based on their own clinical experience, referring to the clinical experience of their mentors, and what doctors wrote about in their clinical experience. Just as horse trainers before 2015 made decisions about horse training systems implemented by equestrian organisations engaging in specific equestrian disciplines (horse sports) and educational institutions, horse trainers, instructors, and coaches made decisions for horses also based on their experience of horse training. Some practitioners coupled certification with experience, provided by the organisation that later became a charity, the British Horse Society (BHS). Like doctors, BHS coaches, trainers, and riders were taught traditional equitation. Only in more recent years have they worked to debunk its once positively regarded reputation for having a 'BHS way.' Rooted in military equitation (functional for taking men into battle), equestrian coach education, at least in the United Kingdom, was founded on the traditions of military horse training. The origins of UK horse training stem from the experiences of military riders training horses for control and obedience, successfully transporting servicemen (and weapons) into and surviving combat. Horse sports did not arrive officially (horses were used in 680 BCE in the Ancient Olympic Games) until 1900 at the Paris Olympics, but then nothing until the permanent inclusion of horse sports in 1912 at the Stockholm Games, when dressage, eventing, and show jumping were introduced under FEI rules.

Like doctors, horse trainers apply experience to training horses; what trainers do not do is refer to EBE literature (peer-reviewed) in the way doctors do now. If your experience of unwanted behaviours (bucking, spinning, rearing, bolting) is that of the horse being 'naughty,' 'not respecting you,' or 'needing more submission' in the dressage test, as opposed to scared, confused, or a horse experiencing pain, experience is misleading. We have historical data showing high rates of equine euthanasia for 'bad behaviour' (Odberg & Boussou, 1999). Horses in history have literally paid with their lives for equestrians placing experience at the centre of the attributes of a horse trainer. The story sometimes still heard to this day is: 'even the professional rider we sent her to said she had spent too long behaving this way (naughty), so if a professional can't help her change (naughty horse), we have no choice but to make the heart-breaking decision of having her euthanised.'

Fortunately, as EBM grew (doing double-blinded, randomised trials), it turned out that the clinical experience of giving patients antiarrhythmic drugs killed people rather than saved them. Clinicians used to encourage perimenopausal patients to use hormone replacement therapy to reduce myocardial infarction until trials showed it increased the risk. There are instances in medicine where benefits doctors thought were present, due to biological arguments, were not. In some instances, biological logic (experience) was harming patients. According to EBM, it is not that there is no role for experience in practising medicine, but where doctors thought biology worked a certain way, it turns out from extensive trials that the biology is complicated and counterintuitive. By zooming out and learning from the field of medicine and the education of doctors applying EBM at the beginning doctors too were defensive. However, hostility towards EBM also helped progress the EBM movement. How? By telling senior doctors, education authorities, and effective clinical practitioners, 'Sorry guys, your training didn't include a certain aspect that really turns out to be quite important. And the way you've been making decisions really has its limitations.' Doctors were not thrilled to get that message, reportedly causing a lot of hostility. However, at the same time, leadership in North America was quick to uptake the EBM movement, publishing EBM in the world's leading journal, followed just 2 years later by the American College of Physicians stating 'in this era of EBM,' demonstrating extraordinary uptake by those at the heart of medical leadership.

Despite hostility on the ground, medical leadership bought into EBM in a big way. No 'light touches,' like Pony Club Australia went 'all in.' In 2018, the leadership team of Pony Club Australia decided to implement throughout its curriculum First Principles (ISES) and the application of equitation science (evidence-based knowledge). In medicine, it quickly adopted its learning of EBM by bringing the principles of EBM into residency programmes. Followed by every textbook, if the book was to be credible, it had to be talking about an evidence-based approach to medical conditions. Mount St John (MSJ) International Dressage Stud, led by visionary Emma-Jane Blundell, made the leadership decision to go 'all-in' back in 2015. Applying First Principles (ISES) by the stud team member, every mare and foal is interacted with using learning theory, shaping and centring the horse's emotional state (affect—Principle 4) in every interaction from the moment a foal is born. Leadership in medicine and horse–human interactions is *doing* what is right for horses *and* people. Providing horses and equestrians with the opportunity to live a 'good life' through identifying negative experiences, helping horses resolve negative experiences, and providing opportunities for positive experiences is how we earn public trust.

ANTHROPOMORPHISM AND REFLEXIVITY

As we have explored, reflexivity is a practice of engaging with questions about our own positionality and different identities how our positionality affects our thinking,interpretation of epistemology (how we know what we know), and relationships. Critical reflexivity impacts the decisions equestrians make in relation to available knowledge and our unique intersectionality with anthropomorphism. Anthropomorphism is the giving of human characteristics, like talking horses and managing animals as humans; 'today I'm giving my horse a duvet day.' Stories like *Winnie the Pooh* are anthropomorphic; animals talk and act like people. In education and research, scholars have used critical reflexivity as a practice to acknowledge one's positionality and interpretation of what the horse is experiencing. We will never know what it is like for a horse to be a horse, so what is the next best thing? Recruit the latest evidence-based knowledge in the fields of cognition, physiology, ethology, and biomechanics, affording robust inferences of their lived experience. We can also enhance horse welfare by reflecting on our propensity to apply scientific scepticism. Intelligent, honest people can unknowingly deceive themselves and others—a condition of being human. By applying the practice of critical reflexivity, we can buffer our horses from our well-constructed rationalisations, justifications, and narratives, unknowingly leading to self-deception.

LOOKING BACKWARDS: THE CLEVER HANS PHENONOMEN

In the late 1800s, in an era of industrialisation, electricity, and the theory of evolution by natural selection, the world was changing. Thanks to evolution by natural selection, science and medicine were experiencing paradigm shifts, becoming popular topics of conversation. The idea that human intelligence didn't just appear one day, but evolved over time from more primitive, animalistic origins was a fresh concept. The notion that animals all around us, like birds, apes, dogs, and perhaps also horses, were in some way intelligent, by degrees comparable to humans and in ways we had not yet considered, was also a fresh concept. Around the same time, mesmerism, mentalism, and spiritualism had risen in popularity, giving a surge in acceptance of fortune telling, séances, paranormal ouija boards, psychics, and telepathy. Lincoln's wife even held seances in the White House. Professional magicians, escape artists, and other masters of stagecraft like Houdini became enormously popular, and as professional deceivers, they couldn't help but notice the similarities to what they did knowingly and openly as tricks and what many of these so-called mediums and psychics were doing fraudulently and exploitatively. As the 19th century came to a close, a math teacher and amateur horse trainer, Wilhelm von Osten gathered large crowds in Germany with his stallion known as Clever Hans. News of Clever Hans's almost human intelligence wowed people far and wide, so much so that Clever Hans would go on to change psychology forever.

A key element of the Clever Hans demonstration was the fact that Wilhelm von Osten usually used a blackboard, piece of paper, or some other prop with a series of potential answers to a maths problem. Von Osten would show a few wrong answers and a single correct answer to Hans and the crowd. Then, Von Osten would point at these one at a time until Clever Hans pawed his hoof to the ground, indicating this was his selection. Then everyone would gasp and applaud. Wilhelm von Osten would be in delighted. It would all go like this, questions about how to spell certain words, how to divide complex fractions and problems, and Hans would paw a few times. Wow. This horse understands language, the calendar, and maths. Von Osten never charged admission for demonstrations, and the crowds grew large because of this. The word got around as he travelled from town to town, and that's when the German Board of Education got involved.

As magicians and professional deceivers, they couldn't help but notice the similarities to what they did knowingly and openly as tricks and what so called mediums and psychics were doing fraudulently and exploitatively. So they used their fame to demonstrate the trickery involved in all of the above when it was used to scam people. A professional debunking movement emerged. Also, at this time, psychology was entering the scene and becoming an exciting new science, especially with scientists like Sigmund Freud debunking pseudo-scientific claims to do with the mind. Later on, scientists did the same thing to some of his claims. The susceptibility to deception converged, and scientists identified it as one aspect of the human unconscious known as the ideomotor effect. The ideomotor effect is when the unconscious mind, through involuntary movement of muscles, responds to thoughts, expectations, or suggestions and does stuff without our conscious awareness. This results in moving our bodies unconsciously, not only deceiving others but deceiving themselves. Psychologists of the time found that the unconscious and involuntary aspects of the ideomotor effect explained how ouija boards moved during seances. In all these cases, people were unconsciously moving their bodies in ways that appeared to be directed by something outside of them, unaware they were the culprits. Instead, participants attributed the movement of ouija boards to the work of ghosts. Sceptical scientists of Clever Hans formed a panel of 13 people, including a vet, circus manager, and zoologist, and created a series of experiments to be conducted by a German psychologist. Von Osten truly believed in his horse, so he agreed to the experiments. In 1907, all the experiments were completed and thoroughly debunked Clever Hans's intelligence due to double-blind testing.

Intelligence tests:

- No crowd or other people questioned Clever Hans
- Progressively moved the questioners further away from Clever Hans
- Blindfolded the questioner so that the person asking could not see the correct answers and sometimes blindfolded their faces

Hans only gave correct answers when the questioner knew the answer *and* Hans could see the questioner. Hans's accuracy dropped from 50 out of 56 correct answers to 2 out of 35, all the way down to levels equivalent to random chance. Hans was reading the facial expressions and body language of the humans, and the humans were unknowingly, unconsciously slightly altering their faces, eyes, bodies, arms, hands, and even their breathing as they approached, glanced at, noticed, or paid attention to what would be the correct answer in each situation. The crucial finding was that Von Osten had no idea. Unaware he was influencing Clever Hans, Wilhelm von Osten believed he was observing independent agency and intelligence in much the same way people in other situations were attributing their movements to telepathy or the power of the undead. After a year and a half of study, the Commission concluded that although Hans wasn't actually reading spelling or doing the arithmetic, there was no hoax involved. It wasn't fraud because it was a case of *self-deception*. Yes, self-deception led to the deception of others, but it was self-deception that was driving all of it. Clever Hans was so good at classical conditioning, First Principle 7, that even once exposed and scientists attempted to hide their unconscious cues when repeating the demonstration as questioners, even attempting to completely hide any possibility of cueing, Hans picked up something if they knew the answer. Clever Hans picked up on microscopic changes in movements and breathing. In the presence of the correct response versus the incorrect one, Hans read humans. Applying the mode of learning; classical conditioning, Hans changed psychology forever. Now known as the Clever Hans phenomenon, when an animal or person performs a complex task, especially outside their abilities, they are responding to subtle, unintentional cues. The value of practising critical reflexivity is the crucial ingredient in so many other experiments; when someone is cueing, moving their body in some way that seems to lead to an outcome that couldn't possibly be coming from themselves, it can very much be coming from themselves. You can be doing something unconsciously and unknowingly, and as with Clever Hans, your horse is picking up on cues in those situations. Knowledge of this effect, double blinding, became a vital part of psychological experimentation and today is an established safeguard in the scientific process. You only need yourself. You only need self-deception. When studying behaviour, face-to-face contact between the examiner and the examined must be avoided.

HARNESSING CASCADES

Transformational change for horses and the spread of welfare reform amongst equestrians can be understood through the lens of community influence. Drawing on Greg Satell's 'cascades' theory, cultural change, such as adopting a welfare horse sports mindset, often occurs through the power of small, interconnected groups within the equestrian community. These new behaviours proliferate through 'cascades' that spread rapidly across CoPs. These cascades frequently originate from small, loosely connected groups of equestrians united by their dedication to the well-being of horses. When equestrians within one community understand and implement the First Principles (ISES), their influence can extend to individuals in different communities, such as the example given of Pony Club Australia influencing Pony Club Canada and the United Kingdom. This interconnectedness between communities allows the First Principles to cascade into adjacent groups, facilitating widespread change. The effectiveness of this 'cascades' approach is linked to the structure of the community, where information and new behaviours are observed to spread more effectively through casual acquaintances who bridge different groups, rather than solely within close-knit circles. The equestrian who connects different communities plays a vital role in disseminating the First Principles (ISES), acting as a catalyst for the cascade effect and driving transformational change for horses. This aligns with the concept of a 'welfare horse sports mindset,' which encourages a 'curious mind' and normalises asking 'how?' to understand the horse's lived experience. By recognising and leveraging CoPs, leaders within equestrian organisations can develop strategies that reinforce and reward the adoption of First Principles (ISES) across various equestrian disciplines and communities, both in person and online. This encourages the spread of pragmatic innovation and creative ideas for implementing positive changes for horses. The underlying aim is to cultivate a culture where equestrians collectively strive to leave horses better, a core tenet of the welfare horse sports mindset.

Understanding how new behaviours, such as the application of the First Principles (ISES), spread within equestrian communities requires considering the influence of social networks and interpersonal connections. David McRaney, in his book *How Minds Change: The Surprising Science of Belief, Opinion, and Persuasion* (2022), explores the concept of 'social death.' This refers to the fear or anxiety associated with being ostracised or losing important social connections and one's sense of identity. This fear can be so powerful that it can even outweigh the fear of physical harm, highlighting the fundamental human need for belonging. When a 'how we have always done it' mindset encounters the welfare horse sports mindset, which encourages asking 'If we knew better, would we do better?'—a complex social dynamic emerges. Exploring contradictory evidence regarding traditional horse training methods becomes intertwined with our social identity and the community we identify with. If this exploration risks alienating us from our social group, potentially leading to a form of social death, it becomes clearer why equitation science might have remained on the periphery of horse training for an extended period. The desire to maintain social connectedness and identity within a community can override the acceptance of evidence-based knowledge in horse training, making established beliefs resistant to change. This phenomenon underscores the intricate relationship between individual beliefs and CoPs, and the significant impact these communities have on shaping minds and reinforcing beliefs about horse training, whether evidence-based or not. However, change can occur when the fear of not changing becomes greater than the fear of social ostracism. This is similar to the idea of a 'tipping point,' where it becomes socially risky *not* to adopt a new behaviour. Leaders within equestrian organisations can leverage the understanding of these social dynamics and the role of CoPs to design strategies that encourage the adoption of First Principles (ISES). By cultivating new normalised behaviour—asking 'How?'—and rewarding the spread of evidence-based practices that aim to leave horses better, the potential for widespread, positive change in horse welfare can be realised.

SOCIAL IDENTITY AND EVIDENCE-BASED HORSE TRAINING

The persistence of outdated horse training beliefs is significantly influenced by the strong desire to maintain social connections and one's identity within a community. This need for social belonging can often override the acceptance of evidence-based knowledge, making it challenging to shift away from established practices. However, this phenomenon highlights the powerful role of CoPs in both reinforcing existing beliefs and potentially driving change towards evidence-based horse training. The 'dancing man video tipping point' illustrates this, showing how the fear of social exclusion—the moment when not adopting a new behaviour becomes more socially risky than adopting it, thus avoiding 'social death'—can ultimately lead to widespread change. In the realm of horse welfare, significant catalysts have initiated a shift towards reform. Immense societal changes in attitudes towards horses in sport, coupled with the consistent exposure of equestrian professionals engaging in abuse through video releases, have created momentum for adopting a welfare-focused mindset in horse sports. This has led to a tipping point for the application of evidence-based knowledge, particularly embodied by the First Principles of the

ISES. If organisations accurately interpret this movement towards welfare reform, they will observe a growing embrace of a welfare horse sports mindset within their communities, reflected in the practical application of these First Principles. Embracing these principles can be seen as the initial step in safeguarding individuals from the potential 'social death' associated with resisting these evolving welfare standards.

Facilitating the application of EBE signals among both equestrians and non-equestrians, even those not actively abusing horses, my experience in welfare education reform. A significant factor influencing the adoption of equitation science literacy is the powerful interplay of social identity and the fear of losing social belonging. Recognising this, the facilitation of CoPs becomes a crucial and valued endeavour. Within these supportive environments, learners are guided through the application of the First Principles (ISES), the Five Domains (2020), and the Quality of Life framework. This process involves challenging existing knowledge, exploring personal identities, and navigating the discomfort that comes with releasing outdated beliefs. The shared experience of confronting these challenges, practising curiosity, and embracing new knowledge—the 'now we know better, let me do better' mindset—is vital, as learners support each other through this demanding yet rewarding journey. As individuals integrate this welfare horse sports mindset, they naturally begin to cascade their learning into their diverse CoPs, creating a ripple effect through the modelling of welfare-conscious behaviours. Understanding how evidence-based knowledge, such as the First Principles (ISES), and the practical application of learning theory spread within these social networks highlights the critical role of interpersonal connections in driving meaningful human behaviour change towards improved horse welfare.

Equestrian communities identify education often as *the* solution for improving horse welfare, or at least as the first option that people think is needed. So now let us deep dive into communication strategies for education (know better) and how that is impacting horse welfare outcomes.

Behaviour change is complex, so experts from the field of health have designed a behaviour change wheel in an attempt to solve rising societal challenges stemming from ill health, such as obesity, heart failure, diabetes, and smoking. The wheel starts by identifying the source of the behaviour. There are three origins of a behaviour requiring change: capability, opportunity, and motivation. In the field of human behaviour change, these sources of behaviour have an acronym COM-B and are referred to as the COM-B Model of Human Behaviour Change. Capability, opportunity, and motivation-behaviour. Education has been flagged frequently by equestrians in different communities in attempts to solve equine welfare challenges. The main source of behaviour requiring change, as identified by equestrians (education), is therefore is *capability*. Understanding capability as the source of the absence of knowledge of how horses learn results in engaging intervention functions, which can be a physical or psychological capability. If change needs to happen for psychological capability (know how horses learn), then knowledge via an online course has impact on psychological capability. Physical capability is the *application* of learning theory (First Principles, ISES); thus, the intervention is skills training (in-person). When there is a miscalculation about the source of the behaviour we want to change (knowledge gaps concerning how horses learn), and we offer an online course (knowledge) alongside an in-person course (skills), while the actual source is the motivation to change, behavioural change is derailed, impacting horse welfare outcomes. The 'This is how we have always done it' mindset resists change. By understanding that behaviour change is a wheel, and the source of a behaviour sits in the centre, we can then zoom out and view all the available interventions and policy categories (see Figures 4.1A and 4.1B).

Identifying the source signposts equestrians to the appropriate intervention. Once a source is correctly matched, a total of nine levers can be operationalised and measured to assess the impact of a behavioural change intervention.

The Behaviour Change Wheel

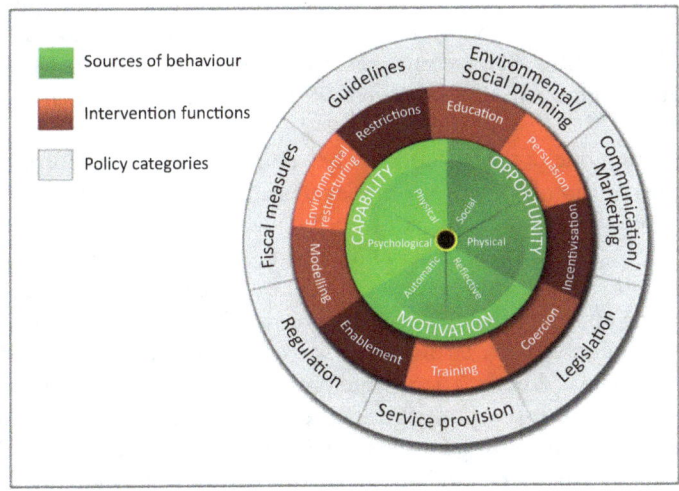

Figure 4.1A The Behaviour Change Wheel

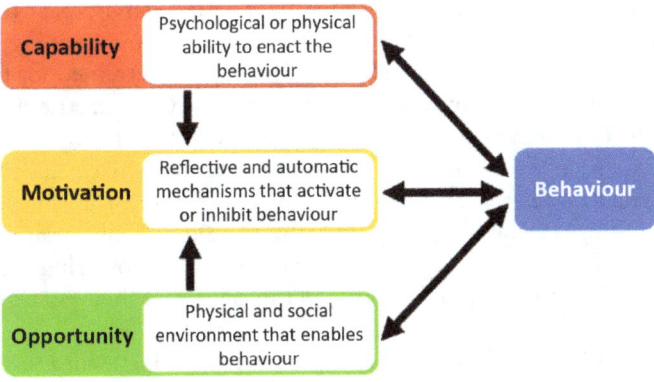

Figure 4.1B The COM-B System—A Framework for Understanding Behaviour

DEEP DIVE

In 2024, I delivered a series of webinars titled 'Elevating Equestrian Practices: An Integrated Approach to Future-Proofing Riding Schools' for the ABRS+, a focused series contributing updated knowledge in equine welfare and ethics for the grassroots equestrian community responsible for the well-being of horses and riders. You can find each webinar recording on the ABRS+ YouTube channel at *https://www.youtube.com/@ABRS-plus/videos*.

Identifying one of three sources of behaviour signposts an appropriate intervention. A total of nine levers can be operationalised and measured to assess the impact of a behavioural change intervention. A team of Swedish researchers investigated the motivation of dairy farmers to engage in veterinary herd health management. Could Swedish vets receive knowledge and training (workshops, role plays) to communicate with farmers in a way that does not evoke resistance to change? Veterinarians were trained in the skill of MI (intervention) and resisted the 'righting reflex' (telling, directing, and conniving). Educating *at* farmers to communicate about herd health management had not worked, and so the Swedish team applied a knowledge and training communication intervention. Whilst passionate about improving animal welfare, loss of control over change creates a backfire effect and we double down on the unwanted behaviour. If we are choosing to educate *at* equestrians, expect to evoke resistance to change. MI is a communication methodology proven to stimulate behaviour change. The vets were trained for 6 months in an MI programme that included role plays to practise the new skill. Researchers found MI was perceived as useful by vets, improving their communication skills in veterinary herd health management. Other areas in which vets were trained included understanding the clients' motivations, that is, values, needs, desires, barriers, and active listening (show empathy), and empowering the clients (work with them to feel confident in achieving their goals).

Working with UK riding schools, applying knowledge (education) and skills (training) application of First Principles (ISES), the human behaviour change model COM-B identifies the source of behaviour (horse training practices) as capability and thus intervention education and training. Yet let us take a brief look back at 1974 when psychologists believed that people, for the most part, are rational optimisers, making judgements and decisions that best maximise the potential of the outcomes under their control. However, two psychologists, Daniel Kahneman and Amos Tversky, worked together and changed the way we think about the way we think. Back then, psychologists didn't see human error as something that could be predicted, something that was systematic, but this changed in 1974 when Amos Tversky and Daniel Kahneman created a paradigm shift beyond psychology, changing the way all of us see ourselves. The mental shortcut or heuristic (an easier path, cognitively) emerged in psychology, which basically means instead of making decisions as maximisers, because we have limited brain power, limited time, and limited patience, we are satisfiers, not maximisers. Humans will settle for good enough in all sorts of situations. When deciding where to eat, we often go with the easiest option, the nearest place. Not because that's our favourite, or it's particularly good, but because it would be a hassle to open our phone, search for the best food in town, talk to people around us, and ask for what they think is the best place. Then take the time and effort to travel to where everyone says you should go, where your phone says you should go, and where the search engines say you should go. It's easier to just go to the closest place, even if that's a work cafe. All of our little decisions all day long are often heuristical (short cuts), but when we look at the entirety of our decision-making over the course of a day, a week, a month, or a year, it can seem like we have been making decisions thoughtfully and carefully and maximising, doing the best thing you could possibly do in that situation. We are bypassing the laborious task of calculating probabilities, and our brain is looking for clues that fit a stereotype instead, like a lazy detective jumping to conclusions because it's easier.

These two psychologists uncovered this psychological paradigm in the 1970s, known as the psychology of single questions. Asking questions in just the right way, certain questions reliably produce errors in decision-making and judgement, providing new insights into human behaviour at that time. Their findings upended common assumptions in the field of economics at that time. It is reported that it upended economics itself, leading to a whole new field: *behavioural economics*. Kahneman received the Nobel Prize in 2002, and Tversky had unfortunately passed away in 1996. Kahneman and Tversky changed the way we think about how we think.

Before recruiting a human behavioural change model, do we start from the place of thinking we are maximisers (make decisions to do better for horses, because we now know better), when in fact we are satisfiers? The 'this is how we have always done it' mindset resists education, skills, and communication interventions. It reflects the 'know better, do the same' attitude towards horses. When jumping in to solve horse welfare problems with more/better education (knowledge), skills (training), and communication interventions (MI), there is an absence of understanding the equestrian mindset. The story of Kahneman and Tversky generated a paradigm shift so powerful that it reached far outside their respective fields—economics and psychology—changing the way all of us see ourselves. This is fascinating and serves as a lesson for today's equine welfare reform. Similar learnings apply to the separate fields of equine performance and animal welfare, giving birth to a new field of research—equine welfare performance. Like behavioural economics grounded in horse welfare and horse sports, we want to evidence that our horses had a life worth living. And for equestrians, imagining what our future holds, we will see how many equestrians we've inspired to *do* welfare horse sports.

By discussing the human behaviour change wheel and the absence of self-leadership in the process of human behaviour change, this is a point of difference for welfare horse sports and horse sports with welfare. *Doing* (not professing) transformational change for horses' mental and physical well-being is cascading the welfare horse sports mindset. Diverse perspectives (formed by unique experiences, beliefs, and social circles) are not just nice to have inside a community, it is this diversity that significantly impacts the dissemination of evidence-based knowledge and pragmatic innovations for positive horse welfare, facilitated in the community. Instead of fearing points of differences and dissenting civil conversations (read here: *https://therulesofcivilconversation.org*), allow diversity of thought to give birth to creative solutions, a foundation for pragmatic innovation to flourish. Dr Pippa Grange, in her book *Fear Less: How to Win at Life Without Losing Yourself* (2020), reframes fear. By cultivating CoPs that prioritise well-being and achievements meaningful to horses (competency, resiliency, and flourishing as a horse, not humans) we lead a more fulfilling and fearless life. The point of learning is not to affirm what we currently believe It's to evolve what we believe." "Great minds challenge each other to think again. Differences aren't division, they are the harmony that helps us grow" - Adam Grant.

 REFLEXIVE PRAXIS

Watch the interview with Dr Pippa Grange here: *https://www.youtube.com/watch?v=7s4SqwluOzw*.

Dr Grange works as a cultural coach across elite sports. Head of People and Team Development at the UK Football Association, in 2018, Dr Grange worked closely with the England football team for the World Cup, a performance that inspired a new narrative of competing with less fear and ego, iterating relationships at the heart of everything, and the antidote to fear.

> Change is hard because people overestimate the value of what they have and underestimate the value of what they might gain.
>
> Belasco and Stayer (1994)

CHECK THE PULSE

> The FEI acknowledges the invaluable role that the media and research expertise play in highlighting issues within our sport. Without responsible journalism and the work of the academic community around the world, equestrian sport would not have evolved to the level of professionalism and welfare standards we see today. We respect the diversity of voices within the equestrian community, including those who challenge our practices and push us to continuously improve. These voices serve as our mirror, aiding us in our daily work of ensuring the welfare of the horse remains our top priority.
>
> FEI Media Statement (19 March 2025)

A welfare horse sports mentality is finding your role in being the voice for the horse. When we choose to stand up for horses over protecting our own interests, doing what is right and erring on the side of the horse is the path to *earning* public trust. However, there are different roles and ways of speaking up for horses and asking for accountability. It always takes courage. Every time I upstand for a horse, my heart pounds and my whole body becomes aroused; I'm *in* the fear. With practice, I have learnt to navigate my fear with a question. My first sentence on behalf of the horse is a question. Serving to interrupt the practice and therefore the suffering for the horse, my question to the rider/coach/steward serves as an interruption for the horse, followed immediately by an authentic curiosity. My tone is curious, not judgemental. I have done the work on 'speaking truth to power' (even worked on my voice shaking), the courage to speak up embedded in my deep curiosity, with the interrupting question being my 'north star' for stepping into courage, in spite of the fear I am feeling. I have learnt that courage is *never* without fear. Courage is feeling the fear and still asking a question. Any question I can think of at that time. Afterwards, I often come up with a better question (reflexive praxis), but my courage practice is asking a question and serving that horse (interrupting the harm in that moment). The best thing about practising courage, especially when I think I can't? Courage expands the more you practise. Just like a muscle. The more you use it, the bigger it gets. When Genevieve Hansen, an off-duty Minneapolis firefighter at the scene of George Floyd's murder, repeatedly asked police officers to check Floyd's pulse, the world witnessed upstanding. Officers at the scene did not provide medical assistance. Asking a question does not even slightly control the outcome. Every time, I literally have no idea how the next few minutes will unfold. None. But it is my shape and shade of horse advocacy. My active stand up for the voiceless, living my life's purpose to leave horses better. And that is why I sleep well at night. Living into our values always takes courage.

So how do we decide when to stand up and when to protect ourselves? As the FEI states, 'we respect the diversity of voices within the equestrian community, including those who challenge our practices and push us to continuously improve,' and 'the welfare of the horse remains our top priority.' According to the FEI, upstanding is valued, even integral to horse welfare horse sports. So the next time you ask is this an upstanding moment? Remember what the FEI values... 'These voices serve as our mirror.'

It is important to discuss the reality of horse advocacy. This section explores upstanding for horses, understanding that your voice for the horse is valued by the FEI, and what the the real challenges that horse advocates face are. There are different forms and shapes of horse advocacy, and Bill Moyer explains that in all *movements for change* in welfare horse sports, there are different roles. Bill Moyer names the four roles as the citizen, the reformer, the rebel, and the change agent. For each of these roles, he talks about what makes someone effective and ineffective. A 'citizen' is someone who is grounded in equestrianism, who promotes positive horse sports, widely held values like freedom to choose and absence of horse suffering, and protects against extremism of all kinds because they solidly believe in sustaining horse sports and contributing to them. A 'reformer' is someone who uses official channels to make change. So that involves thinking like lobbyists, setting up petitions, lobbying and monitoring that success, gathering data and information, and making use of those channels. A 'change agent' is someone who uses people power. They educate, bring in other 'citizens,' and create and help to support more 'citizens.' They are often engaged with grassroots movements and look at long-term alternatives and paradigm shifts. The fourth type is the 'rebel,' who stands up and speaks out, using non-violent disruption and civil disobedience. They are the people who put the problems in the public spotlight and take risks, being brave and courageous.

 REFLEXIVE PRAXIS

What role does your voice play in horse welfare? As the FEI states, 'the welfare of the horse remains our top priority...*these voices serve as our mirror.*'

The different roles for change offer equestrians diversity and language. A positive disruptor is not a negative role, but a different voice serving horses, acting as a mirror for the FEI. A welfare horse sports mindset offers an enhancement to the nonchalant 'this is how we have always done it' mindset, despite having evidence-based knowledge and practices (First Principles, ISES) for training horses, and because animals matter (sentient beings), horses' well-being matters and so to does providing opportunities for positive experiences. Each quality of Bill Moyers's movement of change is creatively (and postively) disrupting the equestrian status quo. Welfare horse sports is not a one-size-fits-all approach to renovating horse sports (keep what is serving horse well-being, change what is not). I see myself in all of the roles. Like being a responsible horse owner, we have different roles throughout their life cycle whilst always attending to the Five Domains (2020) and the model's three central tenets: identify negative experiences, resolve negative experiences for the horse, and provide opportunities for positive experiences.

Active comes from the Latin word *actus*, meaning a doing. The word activism originated in the mid-14th century from the word actus, and means activity. Think proactive, energetic, and lively. In the 1900s, the word became political, but the early development of activism was advocating for energetic action. When Rosa Parks stood up (sat down) and Martin Luther King spoke out, both took action. Yet in today's society, there is a large crackdown in the United Kingdom on activist behaviour as people respond with non-violent disobedience to place problems in the spotlight. Currently, there is a UK government clampdown on disruptive protesters, with legislation being passed on 5 April 2024 preventing repeated disruption at protests.

The climate and biodiversity crisis had people actively talking about what was coming in the 1970s; then those predictions turned into what countries are living through today. For the climate change activists, it felt and feels like no one is listening.

 REFLEXIVE PRAXIS

Rollkur, tight nosebands, soring, whipping, and training methods rooted in creating fear and blind submission—why are they not relics of an equestrian past? Why do you think these abusive equestrian practices are still here? How do some equestrian communities excuse suffering? How is it that with growing public awareness, research, and the latest evidence-based knowledge, still today we have governing bodies, organisations, and equestrian professionals defending horse abuse? If the public can look at a suffering animal and recognise cruelty, why do equestrians deny, dismiss, or enable suffering?

In this infodemic era, is ignorance an excuse? Or is it time to advocate for horses by questioning outdated and unethical practices, standing up for horses in our different roles, and refusing to be complicit in horse suffering? If we act for horses, not for prestige, power, and profit, can we speak collectively? We can expand equestrian courage by *doing* welfare horse sports.

As with the story of the great turning, we can all make a difference to the well-being of horses and the sustainability of horse sports, creating an equestrian world that is healthier for everyone. If we end up going into the great unravelling story, where ineffective equestrians come from a place of fear and distress, this ineffective activism does not bring equestrians along due to extremism. Advocating for horses and bringing equestrians with us, calling in, not out, and modelling enhanced well-being of horses, can begin 'the great turning' for horses.

Finding common ground, the horse, benefits the horse. As wellbeing elevates for horses, Equestrians' own wellness also escalates. Welfare horse sports is a mindset honouring different equestrian journeys and also the equestrian journeys we take with each other, the shared mindset of welfare horse sports. It is about not looking or turning away from the absence of opportunities for horses to thrive. Thinking only about symptoms; blue tongues, hyper-flexed necks, and tight nosebands makes equestrians scared of activism, of becoming activists, and encountering activists, choosing instead to protect ourselves. We want to look after ourselves, ensuring we are safe. Positive horse welfare is an opportunity, an invitation for equestrians to serve the horse. For decades, horses have served us in horse sports. How well are they faring now? When horses are unsafe (blue tongues, hyper-flexed necks, and tight nosebands), that impacts us. It is not okay. Equestrians are part of something more than themselves. That is why welfare horse sports stands for erring on the side of the horse, the horse's fairing well (identifying and resolving negative experiences) intersecting with well-being expansion (opportunities for positive experiences).

> The arc of the moral universe is long, but it bends towards (horse) justice.
>
> King Jr (1965)

My role as a change agent teaches me grace. Every equestrian is a micro-ecosystem of uncertainty and knowing. With much practice, I am learning we do not know or cannot know what comes up for someone when standing up for a horse. When an equestrian responds with fear, often outward anger, this is a 'welfare horse sports blockade.' Meeting anger with compassion and not the anger or the fear they are generating is hard. Because change comes from a place of knowing inside us, deep down, that things need to change for the horse. People block change because they are terrified of what will happen to them, their job, their profit, their sources of revenue. Equestrians working to hold up the status quo are fearful. Feeling threatened in some way, they are not able to live their current life, experience freedom to keep doing what they do, or the stability they feel is at risk if horses experience positive welfare. But equestrians can pivot. We can change the story of the lived experience of sport horses by holding mirrors up to the lived experience of the horse *and* creating windows. That notion of being the change. Individual leadership. Recognising that advocating for horses is how and who we are, not what we expect from what we do, echoing Mahatma Gandhi's 'Be the change,' from taking the bars down in a stable block so horses experience physical touch in the stables to racecourses providing racehorses the opportunity to experience rolling in a sand bath the morning of a race.

TRIPLE WELL-BEING: EQUINE, EQUESTRIAN, AND ENVIRONMENT: LEADERSHIP

> Everyone is responsible for the future of equestrians sports. All equestrians need to optimise and prioritise equine welfare and be seen to be doing so.
>
> Waran (November 2023)

The 2024 Australian Government National Statement on Animal Welfare (*https://www.agriculture.gov.au/sites/default/files/documents/national-statement-on-animal-welfare.pdf*):

> Australia achieves good animal welfare outcomes through the development and adoption of animal welfare practices and standards that are underpinned by science and evidence.
>
> We acknowledge the interconnectedness of animal welfare, human well-being, animal and human health, the environment and climate change, biosecurity, and Australia's socioeconomic sustainability and prosperity.
>
> We acknowledge the need to *build trusted relationships* and sustainable continuous improvement across the animal welfare system to progress the vision.
>
> We acknowledge the spiritual connection that First Nations people have with animals, and the importance of partnering with First Nations people to incorporate their teachings and perspectives on protecting and caring for all living beings. We acknowledge the need to continue to develop Australia's capability to support the animal welfare system.

Safeguarding horses is our collective responsibility, but we are currently failing. From blue tongues, tight nosebands, and training systems showcasing whipping, hyperflexion, and control, to lay experts filling our social media feeds, this is directly impacting horse welfare.

PROFESSOR NATALIE WARAN, CHAIR OF THE EQUINE ETHICS AND WELL-BEING COMMISSION (2022–2024)

Culture advances, norms, and standards of behaviour change, and those lanes of expectations change. EBM is today normalised because doctors provided positive endorsement and reform happened. Leadership is often misunderstood as the ability to direct or dictate—but leadership in 2025 is about *influence*. Influence is relational. It's about building connection, *trust*, and understanding. Influence inspires people to follow, not because they have to, but because they *want* to. People follow leaders who show the way *and* connect with them as human beings. The FEI told journalists (18 March 2025) there had never been any intention to use the new noseband tightness tool on every horse at every FEI show, following pushback to the recently announced protocols for the new measuring tool. The FEI chose to leave testing of every horse up to the organisers of a show, stating, 'While organisers can test every horse if they wish, the protocols for each discipline are geared towards random or targeted testing before and/or after the ride.' Journalist Pippa Cookson reported 'widespread consternation' on reading the finalised FEI protocol, as equestrians felt the FEI 'bowed to pressure from riders.'

GOING FIRST

We all possess the ability to lead. To see the humanity in others as we help them to see the humanity in us, to act as agents of change for the horse. Leadership is inclusive, action-orientated, and people-centred. Our influence is how a welfare horse sports mindset activates enhanced welfare outcomes in sport. Control is the delivery of short-term. Control does not create *followership*. Control is not a sustainable way to lead. It is transactional: do this, get that. Influence, however, is *transformational*. Influence inspires action, commitment, and long-term loyalty because it's rooted in the behaviours of *trust*. Leaders go first, advocating for horses, erring on the *side of the horse*, and building trust through *vulnerability*. As Simon Sinek explains, leaders are anyone willing to be vulnerable and transparent. Everyone is responsible for the future of equestrian sports. We need to optimise and prioritise equine welfare *and* be seen to be doing so. That is not achieved through controlling the narrative, offering up overgeneralisation of 'happy horses' and 'partnerships' whilst denying, minimising, and invalidating stress, pain, and conflict behaviours in horses, but by sharing our challenges, failures, and fears. Individual leaders demonstrate authenticity, building and earning trust with equestrians and non-equestrians (public). A leader isn't the one on the yard who has all the answers. A welfare horse sports leader is the one who goes first (modelling a welfare horse sports mindset) with empathy because they have done it (see Figure 4.2).

Figure 4.2 Embracing Vulnerability in Our Leadership of Welfare Horse Sports

Great leaders don't see themselves as great; they see themselves as human.

Sinek, 2010

Vulnerable leaders spur innovation. Research by Dr Brené Brown indicates that leaders who embrace vulnerability cultivate *trust*, spur innovation, and nurture an environment of creativity and development. Humble leaders improve a community's performance. Leaders who embrace humility build psychologically safe spaces (more about this later). According to leadership research, leaders should communicate more of what they don't know, going first show that it is ok not to have all the answers, which is why the welfare horse sports mindset collaborates, co-creates, and cares for each other.

At great personal cost to me, I learnt that having hard conversations is not the hardest thing. Abandoning yourself is far worse. As I lay awake watching a coach instructing her student to repeatedly whip her horse for refusing a jump, I realised I had not only let this horse and a 13-year-old girl down, but mostly I had let myself down. Where was my individual leadership? Where was I going first? This event is an important stain in my career, and one I never want to remove. It is part of my history, reminding me I can have difficult conversations. My stain quietly speaks to me when I am unsure if I can show up and do the hard thing. For my equestrian journey, I have to have the difficult conversations; I must go first, individually leading and advocating for the horse and welfare horse sports to really live into my values. My fear of having a hard conversation meant I walked away. What could be worse than walking away from systemic brutalisation? A minor abusing an animal? By walking away I was abandoning my authentic self. Trust must be earned. I met my shadow self as I walked away from systematic brutalisation. Not showing up and standing up for horses does not keep horse sports legitimate, credible, or accountable. Trust is earned through listening, engaging, and demonstrating through *action*, individual leadership, going first, and prioritising horse welfare. Everyone is responsible for the future of equestrian sports. We all need to individually lead, optimising and prioritising horses' physical and mental well-being. When we show up for others, serving horses by helping them satisfy their telos, we show up for ourselves. Advocating for horses gives my life meaning. That means today I show up for horses, especially by engaging in difficult conversations. Helping horses is how we help ourselves.

 REFLEXIVE PRAXIS

14 Leadership Lessons from the Apple TV show *Ted Lasso*:

1. Be authentic
2. Stay coachable
3. See good in others
4. Happiness is a choice
5. Winning is an attitude
6. Have confidence in yourself
7. Optimists take more chances
8. Everyone differs from everyone else
9. Courage is the willingness to attempt
10. Vulnerability is a strength, not a weakness
11. Doing the right thing is never the wrong thing
12. Be curious, not judgmental
13. Be a goldfish—don't allow one bad deed define who you are. In less than 10 seconds, forget about it like a goldfish
14. Barbecue sauce. That moment when you know it's all going to work out

BECOMING DESPACITO

In January 2017, a song called 'Despacito' by Fonsi, Erika Ender, and Daddy Yankee was heard on the radio, in pubs, and homes. Within 6 months of release, it became at that time, the most streamed song of all time, with over 4.6 billion streams. Liking 'Despacito' more now than the first time you heard it is due to the psychological phenomenon known as 'the mere-exposure effect' or 'familiarity principle.' The more exposed we are to a song or a set of principles, the more we like it. Frequency and familiarity increase likeability, as is the opposite true. 'Out of sight out of mind' is an accurate phrase in psychological research. The frequency of hearing, liking, and 'becoming Despacito' is the ISES's First Principles of Horse Training. Reviewed and updated three times, ten principles are becoming 'Despacito,' influencing equestrians in the stables/barns, arenas, and riding and pony clubs.

Print the latest version of the poster. Travel with the First Principles everywhere. I have the poster on my office noticeboard, in the glove compartment of the car, and fridge door. If you work at an educational institution such as a college/university or a riding school or livery yard, take the First Principles (ISES) poster to a printer and get them to blow it up to create as a sign for students/clients to see during their riding lessons.

 REFLEXIVE PRAXIS

Print out the First Principles *checklist* https://www.equitationscience.com/ises-training-principles when you go to observe clinics.

Gift the First Principles poster to clients/friends/coaches. Pay attention to all the principles throughout your own training; for example Principle 4 is to pay attention/attune to how the horse feels during your interactions. Scan a lesson/coach for the First Principles (ISES). Just because a coach believes the horse is 'happy' does not make it true. Find evidence supporting any claims about the horse's emotional state made by coaches, riders, and organisations. Like the song 'Despacito,' bring First Principles into all your equestrian communities. Practice the First Principles (evidence-based equitation, EBE). All horse training methods will engage with some of the ten principles. Scanning for evidence of some of the First Principles is better than none. Like 'Despacito' and the 'mere-exposure effect,' familiarise yourself and frequently refer through your horse training to the First Principles. Number and name each principle in practice. Frequency and familiarity increase likeability. A welfare horse sports mindset is increasing the likeability of First Principles (ISES) because doing welfare horse sports is doing EBE.

What is *in* your 'vault' of knowledge? Is the knowledge in your vault standing up to scientific testing? If your vault still contains the application of additional punishment (whipping, punching to suppress an unwanted behaviour) and dominance theory (horse needs to respect you), let them go. Remove them from your Welfare Horse Sports 'vault of knowledge.' Punishment is outdated. Dominance theory is disproved. We are in times of massive moral social movement. Our duty to safeguard horses demands that when we know better, we DO better.

DEEP DIVE: THE EQUITATION SCIENCE SMOOTHIE

What 'ingredients' do you have in your smoothie to safely, effectively, and ethically train different horses? Fill out your equitation science smoothie, one for each horse you train. If they are unfamiliar, the ingredients can be searched online. Every horse has a different learning history, current environment, and genetics (breed/parentage), so meeting the needs of all horses is more about what 'training tools' you have in your horse training 'toolkit' combined with the skill of shaping at the speed of each individual horse. Do you have a 'signature smoothie'? Review the 'ingredients' that I use to train depending on the individual horse and what I am shaping (see Figure 4.3).

Figure 4.3 The Equitation Science Smoothie

LIVING INTO OUR VALUES

Advocating for horses means listening and having hard conversations. Before we can do hard things, we need to know our values. Once we know our values, we can work at living into them.

 REFLEXIVE PRAXIS

Visit the Clearer Thinking website (*https://www.clearerthinking.org/tools/the-intrinsic-values-test*) to find out your beliefs.

Think about what you deeply, fundamentally value, then examine how you show up for your horse and reflect on how your mindset aligns with those values. Distinguish between your intrinsic values, regardless of where they came from, such as genetics, upbringing or culture, versus when you're living by others' values. Respect all your intrinsic values, even if you recognise that a value comes from culture; everything about us comes from somewhere. We didn't self-create our values. However, differentiate between not having a value, while your friends or parents have that value, and so you are not trying to lean into a value that is not your own; it's someone else's value. A welfare horse sports mindset is the intrinsic value of 'knowing'; learning and truth. Reflect on the value of knowing (learning and truth) (clearerthinking.org) and why equestrians collectively suggest education as the solution to what World Horse Welfare (2023) references as biggest threat to equestrianism in the western world: social licence to operate (public trust).

 REFLEXIVE PRAXIS

Read through the list of values and complete the activity. Now that you know your top two values, the hard work can begin. Values do tweak over time; however, your core values stand the test of time. In 2000, I wrote *Horse Welfare from Inside Out: A*

Values Activity:

Accountability	Confidence	Financial stability	Humor
Achievement	Connection	Forgiveness	Inclusion
Adaptability	Contentment	Freedom	Independence
Adventure	Contribution	Friendship	Initiative
Altruism	Cooperation	Fun	Intuition
Ambition	Courage	Future generations	Job security
Authenticity	Curiosity	Generosity	Joy
Balance	Dignity	Giving back	Justice
Beauty	Diversity	Grace	Kindness
Being the best	Environment	Gratitude	Knowledge
Belonging	Efficiency	Growth	Leadership
Career	Equality	Harmony	Learning
Caring	Ethics	Health	Legacy
Collaboration	Excellence	Home	Leisure
Commitment	Fairness	Honesty	Love
Compassion	Faith	Hope	Loyalty
Competence	Family ©equicoach.life	Humility	Making a difference

Figure 4.4 The Values Activity

Roadmap for the 21st Equestrian Coach and how to navigate social licence to operate (legitimacy, transparency, communication, and trust). My same two core values 5 years later are freedom (agency to choose my response) and curiosity (keep asking 'what am I missing?' and, as part of curiosity, attend to actively listening to answers). I have been leaning into my values wholeheartedly these past 5 years. Our values are my 'north star,' providing my life with purpose, in travelling along my equestrian journey and my path towards living a meaningful life (see Figures 4.4 and 4.5).

PRACTISING OVER PROFESSING

What does living into your two core values mean to you? Earning public trust shines a spotlight on our values. Choosing our values over what is easy takes courage and practice. Acknowledging my own disconnect from the person I wanted to be and my (in)actions was necessary. I had to physically walk away from a horse who needed my help in order to get back to my authentic self. Jay and Grant call this 'dynamic authenticity.' By living into my values

Values

Nature	Risk taking	Travel	
Openness	Safety	Trust	1. Circle 15 values you relate to
Optimism	Security	Truth	
Order	Self-discipline	Understanding	2. Choose 2 central values from these that fuel the others
Parenting	Self-expression	Uniqueness	
Patience	Self-respect	Usefulness	
Patriotism	Serenity	Vision	
Peace	Service	Vulnerability	
Perseverance	Simplicity	Wealth	They are values that fill you with a feeling of purpose
Personal fulfilment	Spirituality	Well-being	
Power	Sportsmanship	Wholeheartedess	
Pride	Stewardship	Wisdom	
Recognition	Success		
Reliability	Teamwork		As you read them you should feel deep resonance & self identification
Resourcefulness	Thrift		
Respect	Time		
Responsibilty	Tradition	©equiocoach.life	

Figure 4.5 The Values Activity

and doing the work to be who I want to become, every decision of every day takes practice and lots and lots of mistakes. A welfare horse sports mindset practices balancing the best self-leadership we can today whilst doing the reflexive work (studying our mistakes) so we can do and become better tomorrow. The difficult conversations will always be difficult. But my ability to show up and step into my leadership (courage to go first) and have the difficult conversation is a mindset shift. From professing horse sports *with* welfare to *doing* welfare horse sports.

DEEP DIVE JOURNALING

Who is someone who knows your values and supports your efforts to live into them?

What does support from this person look like? What can you do as an act of self-compassion to support yourself in the hard work of living into your values? What are your early indicators or signs that you are living outside your values? What does it feel like when you are living into your values? How does living into your two key values shape the way you give and receive feedback?

LEANING INTO VALUES

> Real commitment to prioritising equine welfare, is demonstrated not just by what we say, but by what we do
>
> Professor Natalie Waran, as cited in Lesté-Lasserre (2023)

There is no better measure of values than how we spend our time. Much of what we tell ourselves is different from what we actually do. Accounting for how we spend our time, our minutes, is the hard truth of our values. Take a look at all the actions that serve your priorities. We can tell ourselves all sorts of stories to rationalise how we spend our days. I replay these words; future you needs current you to do work. If you find yourself judging the physical or mental fitness of others, or both, research shows this is where your own deficits in self-worth sit. Shaming others is more about the shamer's own feelings than about improving horse welfare.

 ## REFLEXIVE PRAXIS

How do you live a typical day? Of course, some days are consumed by a short-term crisis or some other indulgence, but it is the compound effect over time that changes how we feel towards our horses. The net effect over time is how we assess our health and well-being, not the occasional junk food, bottle of wine, or Netflix binge.

Audit your routines *before* they become habits. From checking social media throughout the day to going through the motions of your morning routine, you may forget why you even started doing what you are doing in the first place. Identify all that does not take you closer to *who* you want to become. Ask yourself, will the future you thank the current you? Make changes gradually, over time.

Be aware, as a day draws to a close, of willpower fatigue. Mitigate low willpower by the end of the day by recruiting technology. Set a behavioural prompt to do a new behaviour at a time when you least want to, for example, set a notification to make tea instead of a habitual glass of wine. In horse training, this is known as differential reinforcement—cueing a new behaviour and interrupting an existing, unwanted behaviour. For example, if your horse responds with fear to fireworks, *before* bonfire night (5th November), target train your horse (touch an object) followed by receiving a verbal bridge (click/cluck) and then a reinforcer (nut/scratch)—something your horse finds valuable—so they touch the target more frequently and experience receiving the nut/scratch as a consequence of touching the target. Now expand the duration your horse holds to the target. Do this gradually. Place the target somewhere your horse will return to (if stabled). When the first fireworks start (often a few days before 5th November), use differential reinforcement (train your horse to the stationary target). Changing your horse's emotional state (affect) according to Principle 4 (First Principles, ISES) via differential reinforcement during the experience of fireworks is another observable behaviour to leave your horse better because of your interaction/training.

RESILIENCY BUILDING (HORSE)

Building resilience is working on maintaining mental and emotional stability whilst living in a mild state of disequilibrium. The compromises we all experience throughout life are the path to building resilience. Author Ryan Holiday has coined hardship or discomfort as 'the obstacle is the way.'

Over your horse's lifetime, compromises your horse will experience include veterinary interventions; routine injections, endoscopes, or perhaps life-saving colic surgery and the bonded pair to your horse leaving or dying due to a change of human ownership, injury, box rest, loading into small spaces, or travelling over long distances. It could be one or many. Building resiliency is about horses practising modulation via self-regulation of affect (emotions).

A welfare horse sports mindset creates environmental enrichment ideas, like scratch posts made from broom bristles, rubber curry combs, and other brushes with different bristles for horses to get opportunities to experience different ways a bush feels on different parts of their body. Setting up spaces so your horse explores self-regulation in adversity (ability to chew, i.e., access to forage helping modulate stress) is important. There is evidence of horses having broken a leg while grazing. This is because mastication, or chewing, is not just necessary for calorie intake (subsequent energy created from eating grass); it is vital to their species-specific needs. Therefore, in times of dysregulation, chewing and providing opportunities to forage is *essential*. A resilient horse modulates themselves during adversity; separation from a friend/s, travelling (with friends), travelling (alone/without friends), incorrect use of removal reinforcement (in-hand and under-saddle), and veterinary and farrier interventions. *Resilience building* for horses is providing a horse with the opportunity to practice self-regulation *in* adversity. There are learned responses I train all horses to be competent in:

Resiliency Expansion—Observables

Trained Response	*Physiological Change*
Head down/Head up	Lowers blood pressure, also lowering heart rate
Longer strides	Lowers head and neck—poll same height as wither—lowering head associated with activating parasympathetic nervous system, leading to reductions in heart rate and blood pressure, recruiting relaxation

Changing a horse's emotional state during box rest is about tapping into our overstory of leave horses better, by providing new/varied sensory opportunities; smell, sight, touch, sound, and taste. Environmental enrichment for horses is well documented; music, scents, herbs, scratching posts, foraging toys, and mirrors. Building resiliency is about helping horses experience competence *in* adversity. Anything from needing emergency veterinary intervention, experiencing the loss of a pair bond (through moving or death), to the sights and sounds of crowds at international equestrian events.

Practice expanding your horse's resiliency by offering horses the opportunity to explore and be curious (novel enrichment). Practice investigating your horse's ability to self-regulate their varied emotional responses to environmental stimuli (includes us, because we are in their environment).

RESILIENCY BUILDING (US)

> Between stimulus and response there is a space. In that space is our power to choose our response. In our response lies our growth and our freedom.
>
> Viktor Frankl (1946) From *Man's Search for Meaning*

We get to choose how we respond to all that happens to us.

 REFLEXIVE PRAXIS

155 passengers on the US Airways Flight 1549 experienced the exact same plane crash into the Hudson River on 15 January 2009 (see Figure 4.6).

Figure 4.6 It Is Not What Happens to Us But How We Respond That Matters

What response would have been yours, immediately after surviving the crash?

WHY ISN'T EVERYONE RESILIENT?

Your resiliency journey is unique to you. Those who experience adversity get to practise turning obstacles into opportunities. The through-line of resilient equestrians is openness. Resilient equestrians take in, examine, and process new inputs to training and coaching. Resilient individuals in hard times (like during the 2020 coronavirus pandemic) let go of the things that at the time lack value, adapting and moving on to create or experience the next opportunity. Resilient equestrians react to unexpected, life-disrupting change by accepting it. They then work on a mindset update and convert the adversity into events that ripple new opportunities and events.

LEARNING YOUR LANGUAGE

Journal at whatever time fits you (first thing in the morning and last thing before bed are both optimal times to journal). Share your journaling or keep it to yourself. Start with 10 minutes every day. Journaling at different times of the day helps you to learn more about your story. In the beginning, notice how you talk to yourself. With time, things will change as you learn your language.

'CALLING IN'

Calling in refers to the practice of addressing problematic behaviour that emphasises compassion, accountability, and dialogue, rather than shaming, humiliation, or exclusion. The concept (phrase) was popularised by activist-scholar Loretta J. Ross, and contrasts with 'calling out' by seeking to preserve relationships and facilitate learning.

> Meeting people where they are is an ethical mandate.
>
> Dr Brene Brown (2021)

SHAME: Observables

- Perfectionism
- Favoritism
- Gossiping
- Comparison
- Self-worth
- Harassment
- Power over
- Bullying
- Blaming
- Teasing
- Cover-ups

We cannot shame and belittle equestrians into changing behaviour. No matter how much we want to help horses, we cannot force people to make positive changes by putting them down, threatening them with rejection, humiliating, ridiculing, or belittling them.

From the moment I read my first book by Brené Brown, a shame researcher (see resources), I felt overwhelmed by the frequency and intensity of shaming and shame in equestrianism. I still feel overwhelmed. After many hours exploring the science of shaming and shame resilience, I've leaned into my own experiences as a horse owner, competitor, and coach.

Brown states, 'We do not talk about shame. We experience it, we feel it. We sometimes live with it for an entire lifetime, but we do not talk about it.' Brown highlights that shame is a silent epidemic.

Unlike our recent openness to discuss mental health, shame still remains taboo. Yet shame is universal. That painful wave of emotion washes over in the face of judgement, ridicule, and humiliation. Your horse fails to load, you wear tack that fits your horse but looks different, you ask a question in training and you are ridiculed for having 'no common sense.' I've shamed and been shamed.

How do equestrians break the relentless shame cycle? According to Brown, by being vulnerable. Vulnerability is uncertainty, risk, and emotional exposure, which are also qualities of courage. How we choose to make people feel is as, if not more important than how we make our horses feel.

Making a difference for the horse always starts with us making a difference to how equestrians feel. Shame forces us to put value on what other people think. We lose ourselves in the process of trying. The antidote to shame is connection. To cultivate connections, first we must learn how to build resilience to shame.

Like the compromises a horse will experience throughout a lifetime, it is unrealistic to think the compromises in your own life can and should be avoided. Start by connecting with you. Be your authentic you. Find your true self. From the place of being the real you, reach out and connect to equestrians. When you feel seen, heard, and valued, you have connected to 'your people.'

I have been surrounded by people and never felt more alone. Staying away from making judgements is where our own shame resilience work is needed. My daily self-care connects more of me to me by building minute-by-minute resiliency to my own shame. Shame is all about the fear of disconnection. We are biologically wired for connection; socially, emotionally, and cognitively. When we experience shame, it is likely we are afraid we have exposed our flaws, revealed parts of ourselves that jeopardise our connection to others—that equestrian credibility that makes us worthy of acceptance. Shame is about fear.

EMPATHY BUILDING

We can practise developing our empathic skills. Empathy is not inherent and can be learnt.

Stay out of judgement—equestrians' default is to wade straight in.

Try to understand what emotion an equestrian is articulating. Expand your emotional literacy by getting fluent in the language of feelings. Emotional literacy is critical as it gives you language.

Communicate your understanding of that emotion. Place their emotion verbally 'on the table.' If you get it wrong, be courageous and circle back. This exchange builds connection and trust, a solid foundation towards earning your social licence.

There are times when our practising of empathy will go wrong—the empathetic misses. Times when we might share our vulnerable self (e.g., a struggle or even something that excites us or makes us happy) and we won't feel heard, seen, or understood. Brown describes the empathetic miss as 'a sinking feeling where you feel exposed and sometimes right on the edge of shame.' Following are six of the barriers or empathetic misses that, as equestrians we should learn to recognise so we can meet horse lovers where they actually are. Recognising these gives us opportunities to form circles of connection.

EMPATHY MISSES

1. *Sympathy versus Empathy*: Empathy is feeling with people, sympathy is feeling for them. Empathy fuels connection. Sympathy drives disconnection. 'That's bad, I'm so sorry for you' disconnects, whereas 'me too' says I may not have had the same experience as you, but I know this struggle and you are not alone.
2. *The Gasp and Awe*: Your friend hears your story and feels shame on your behalf. 'I was so embarrassed the other day, it took me an hour to load my horse to take to the vet—I was just so thrilled we did load, eventually!' Your friend blurts out, 'I'd just have died; perhaps you should have been practising earlier so the vet didn't have to wait?' You hope your friend would have said a version of 'I've done that, thought he would load; it took us ages! I later found out he was in so much pain going up the ramp—once we got to the vet! They were so understanding when we eventually arrived.' Instead, you rush in, 'No, it's ok.' Suddenly, you need to make your friend feel better.
3. *The Mighty Fall*: Your friends think of you as a pillar of worthiness and authenticity. They can't help you because they're so let down by your imperfections. Their response to you is, 'I just never expected that from you. When I think of you, I

just don't think of you as the kind of person that posts such a controversial opinion.' You're not experiencing a connection in an empathic way. You're defending yourself to someone because they're disappointed.
4. *The Block and Tackle*: This is when you go to your livery owner and say how you have just mistakenly put your horse in the wrong field and now all the horses are running around, and you need help to catch your horse. The owner says, 'How did you let this happen? What were you thinking?' Your friend looks to blame someone else, 'The new livery put her horse in your field, it's her fault!' I've come to you, the livery owner, because I need your help. Your friend is making it easier on herself by refusing to sit with you in discomfort. Choosing to be annoyed at someone else or to stand in judgement is a huge empathy miss.
5. *The Boots and Shovel*: This is when a equestrian is desperately trying to make things better for you so they can get out of their own discomfort. You go the wrong way in your dressage test. Your friend says, 'You know, it's not that bad. You know you are awesome.' Your friend is not hearing anything you feel and is not connecting to any emotion you are describing.
6. *If You Think That's Bad*: This equestrian confuses connection with the opportunity to one-up you. 'That's nothing, if you think those stomach ulcers look bad, look at these—granular grade 4!' Comparing or competing is not empathy. The most important words we can say to our friends and clients are 'me too. You're not alone.'

 REFLEXIVE PRAXIS

How do you rate your own empathetic skills?

When you think about the six empathy misses, are there any that shut you down?

What emotion comes up for you when sharing your story meets one of those barriers?

How does an empathy miss affect your connection with the person?

Are there one or two responses you typically use that you need to change?

 REFLEXIVE PRAXIS

List as many wins as possible (AMWAP). Every day changes the way you scan and view the world. Daily AMWAPs reshape how you see life, the story you tell yourself, and how you choose to narrate your story. Scan for the wins. Complete your AMWAPs every day for 2 weeks.

POSITIVE REINFORCEMENT IS NOT JUST FOR HORSES!

Shine theory is a practice helping each community member be their best self—and relying on the community's help in return. It is a conscious decision to bring your full self to your community (not letting insecurity or envy ripple inside yourself and your community). Shine theory is a commitment to asking, 'Would we be better as collaborators than competitors?' The answer is almost always *yes*. Shine theory recognises that authentic confidence is infectious, and if someone is tearing you down or targeting you as competition, it's often because they are lacking in confidence and/or support themselves. It's a practice of cultivating a spirit of genuine happiness and excitement for your friends doing well, and being there for them when they aren't. Don't mistake Shine theory for networking. Shine theory is not about helping everyone you meet along your equestrian journey either, because if you're doing it right, it's simply not possible to invest deeply in that many people. Shine theory is intentional. It is accountable. It is relational.

When you positively reinforce people, you are responsible for creating the 'rainbow effect'—it is your reinforcement/praise that

Figure 4.7 Praise Prism

transforms ordinary light to the shining rainbow (see Figure 4.7). When the light re-emerges from the prism, everything is brighter, bigger, and better.

 REFLEXIVE PRAXIS

Watch on YouTube a video of Theodore Roosevelt's 1910 speech 'Man in the Arena.' Daring greatly always means you have faced down moments.

Serving horses and enhancing their well-being is not a zero-sum game. It is not that if my horse is getting closer to having his telos satisfied, your well-being is diminished. The welfare horse sports mindset is what we can add to well-being; horse, human, and environment, a collaborative effort of doing the work together to come up with expanding well-being outcomes for equines, equestrians, and the environment.

Collaboration, and more specifically meaningful collaboration, goes beyond working with others. Meaningful collaborations improve horse welfare. When we work together, we can do even better for horses. I am often asked, how do we know with whom we should collaborate? This is my essential guide to collaboration over competition, dedicated to helping equestrians cultivate a welfare horse sports mindset, where abundance is valued over scarcity, authenticity over perfectionism, and reflective thinking over 'this is how I have always done it.' Reaching out to collaborate starts with who we are; what are our two core values (see Figure 4.4 and Figure 4.5)? Find equestrians with similar shared values to yours. Shared values are the heart of meaningful collaborations. Think of the people, communities, organisations, and events that share your 'overstory' and leave horses better. Get clear on the difference between sharing the same through line and agreeing on everything. A person or place that triggers lonely feelings in you is actually a signpost towards a lack of belonging. Finding a place of belonging can happen in the most unusual of places, including from meaningful collaborations. Belonging is complex. To be 'at home' is to be known, to be loved for who we are, and to share a common sense of common values with others who truly care about us. As we know, we can feel lonely without ever being alone. The foundation of feeling 'at home' is that of genuine connection with others. How do we create authentic human connections?

Be interested. This is also how you become interesting. Ask about how *they* learned their skills/experience. Be mindful of anything *they* would like articles, podcasts, memes, anything you think they will like). Tag them in it. Share insightful comments; this can be about anything of interest to the person you wish to collaborate with. Share their content online and tag, saying why equestrians should read this.

RELATIONAL DYNAMICS

In this digital, agentic AI age, there is no substitute for authentic human connection. The genuine quality of connection is referred by researchers as the power of 'relational energy.' Relational energy is the emotional energy generated or depleted in every social interaction. When we feel good, making a strong connection with another person, we gain cognitive benefits, clarifying our thoughts, improving memory and cognitive performance. We are fired up, energised, and engaged in our work, increasing productivity. Positive connections, fleeting or enduring, are marked by warmth, generosity, and a sense of engagement.

LEAVE HORSES BETTER COMMUNITY OF PRACTICE

When we feel isolated, research shows it leads to less robust health, less enjoyment in life, and less of an ability to collaborate to find solutions. When we feel satisfied with our social connections, we feel safe. When we feel safe, we can think more creatively, and researchers found we also anticipate and experience positive emotions more often, providing long-term physiological benefits, plus immediate psychological uplift. It is that boost in mood that ripples into how others behave towards us, encouraging more creative collaborations, creating a cause and effect cycle back and forth in equestrian CoPs, rippling outward in a widening CoP circle. The degree of social connection that can improve our health and perceived happiness is both simple *and* difficult when we are not in community. Loneliness is not a disease; feeling lonely is like hunger or thirst from time to time. Just as thirst is the prompt reminding us to stay hydrated, loneliness is the prompt reminding us how much we depend on one another. The degree of social connection improves health, with those who live a 'low in loneliness' life being fully available for genuine social interactions.

A CoP, created to leave horses better, is fully contributing to a social relationship. Studies have observed from 'low in loneliness' people that their contribution to a community is through quiet encouragement of those motivated to lead, giving others the benefit of the doubt and are more likely to forgive than those who are 'high in loneliness.'

At the heart of all equestrian communities is our shared connectedness to the horse. Building CoPs to experience social relationships is looking at whole systems thinking, an indigenous, old, big idea. Systems are entities of interrelated elements working together to achieve a purpose. When we look at horse welfare through the systems lens, we see *relationships* rather than separateness. From the DNA in our cells to the individuals in our communities, we are engaged in mutual influence, interdependent, collaborative, and co-operative. Our parts of the whole community work together. In whole systems thinking, processes evolve. In Welfare Horse Sports, CoPs aim to leave horses better, always evolving. We will now review the research into high-performance teams, exploring successful guard-rails for individuals shifting or embracing a welfare horse sports mindset and building healthy CoPs.

GUARD-RAILS FOR COMMUNITIES

In recent longitudinal research, psychological safety (PS) was identified as the number one characteristic of high-performing teams. Especially important in health care and aviation industries, PS has been shown to reduce errors and enhance individual and community learning. We cannot step fully into horse advocacy if we are unsafe to speak up. According to Amy Edmondson (2012), a psychologically safe community is one where it is safe to experiment, collaborate with diverse voices, and do reflective practice, referenced as TEAMing.

Observable behaviours of a *safe* CoP:

1. Speaking Up

    - Asking questions, seeking feedback, and making mistakes
    - Asking for support/help
    - Conversing about experiences, insights, and questions that build new practices
    - Encourage difficult conversations

2. Experimentation

    - Learning from the results of action
    - Curious to learn the impact of your actions on the horse/equestrians

3. Collaboration

    - Cooperative (looks like mutual respect)
    - Shared goals (whole bigger than the sum of individuals)
    - Sharing information

4. Reflection

    - What worked? What didn't and why?

Observable behaviours of an un*safe* CoP:

1. Speaking Up

    - Silence is easier than speaking up
    - 'This is how we have always done it' mindset

2. Experimentation

    - Due to strong social and psychological reactions to mistakes, mistakes seem unacceptable.
    - Fear culture cultivated

3. Collaboration

    - In the absence of collaboration, there is an absence of shared goals and mutual respect
    - TEAMing breaks down

4. Reflection

    - Remains out of date
    - Make preventable mistakes
    - Absence of intelligent mistakes

LEADERSHIP IS A BEHAVIOUR NOT A TITLE

Creating safe communities requires understanding and doing 'framing,' an essential CoP observable. Framing is challenging. My experience of numerous and varied CoPs is a rich tapestry of traditional equitation beliefs and resistance to change attitudes, from a 'light breeze' to 'gale force winds' (everything in between).

I apply framing to lead learning the First Principles (ISES). *How* I frame each of the ten First Principles matters to learning the First Principles. The power of framing lies in my ability to use the most relevant analogy for that community. Edmondson (2012) explains that the power of framing is transformative for communities, making learning accessible and interpreting each principle in a way individuals inside each community can learn. Meeting each equestrian where they are is every CoP manager's mandate.

Observables of psychologically safe communities:

- Equestrians with the *least* status find and share their voice; on yards, inside riding clubs, inside online groups, etc.
- Being our authentic selves liberates others watching. Modelling for others, it is ok to be yourself.
- Safe to learn from each other. Being the learner, not the knower builds a culture of curiosity and lifelong learning.
- Everyone deserves brave and safe spaces to be vulnerable; therefore, you work to create brave, safe spaces for equestrians by modelling horse advocacy through evidence-based knowledge (Five Domains, 2020 and First Principles, ISES).
- Model removing own armour, showing what it looks like to go first; you be vulnerable first.

Boundaries to psychologically safe CoPs:

- **Status**: The most challenging boundary in the creation of a psychologically safe CoP. Lower-power equestrians usually find it hard to speak up. I have experimented with reverse mentoring, effectively lowering the status boundary and creating PS. What is observable about reverse mentoring? The most high-status equestrian/s are mentored by equestrians with an understanding and knowledge of learning theory (First Principles, ISES). When status is a boundary to a psychologically safe CoP, do positive disruption of status by setting up and supporting reverse mentoring.
- **Knowledge**: Differences in experiences, knowledge, expertise, and education. According to Edmondson (2012), knowledge is *the* major boundary to psychologically safe spaces. How

do we make CoPs psychologically safe by overcoming knowledge? Value the journey *every* equestrian is travelling along. When we authentically meet equestrians where they are, we 'call in' ways of knowing. My job as a CoP manager is to help equestrians reflect on *how* they know what they know (epistemology). The most common starting place of equestrian knowing is 'Empirical Knowing' (sensory experience). Knowledge gained through observation, experience, and data from senses is also common knowledge amongst equestrians, referred to as 'Authority Knowing.' This is knowledge (whatever content that is) accepted based on the opinion of someone with status and/or tradition; for equestrians this may be coupled to a person's later stage in the cycle of life. 'Intuition Knowing,' referred to as 'gut feeling' or instinct, is referenced early on by equestrians inside the CoP I am invited to work with. 'Emotional Knowing' is emotion and personal connection, referred to by equestrians as 'feel.' Over time, emotional knowing has been 'up voted' and reinforced through authority knowing, making positive change for horses through 'Evidence-Based Knowing,' a big catch up. A CoP centred on 'Evidence-based Knowing' has a structured approach to testing and validating knowledge, centred on systematic, peer-reviewed, and reproducible knowledge. The scientific process (scientific method) provides structure to investigations to *build* reliable knowledge. It is used in all fields of science, including equitation science, optimising objectivity and reproducibility. In addition to 'evidence-based knowing,' knowledge gained by deduction, reasoning, and logic (does not need sensory input) is called 'Reason and Logic Knowing.' Discussing why a horse training system is ethical/unethical relies on 'Reason and Logic Knowing,' as First Principles explain what the horse is experiencing as evidenced by the behaviour of the horse, including their emotions (Principle 4, ISES). The seventh and final knowing is at the heart of a welfare horse sports mentality, 'Reflective Learning,' knowledge built through reflection. A welfare horse sports mentality recognises that the value of knowledge from what does not work is as important, maybe more so, than learning what does. There is even an Institute for Brilliant Failures (IvBM), founded by Professor Paul Iske in 2015 in the Netherlands, whose point of difference is embracing failure (horse training mistakes), an important learning moment, providing opportunities to gain knowledge. The IvBM challenges society by facilitating and making learning experiences accessible, with opportunities to learn from what went wrong. Well-intentioned actions (love horses) and failing to satisfy a horse's telos or mental domain (Five Domains, 2020) are framed in my CoP as 'brilliant mistakes.' This reinforces brilliant equestrian mistakes with awards, inspired by the activities of the IvBM. Did you know there is an annual one-day conference on six continents to learn from and prepare for failure, in order to grow faster?

FailCon is a conference whose sole purpose is to share failures and embrace mistakes. It shifts mindsets towards embracing mistakes, failures, and face-down moments to accelerate growth. Failing forwards. Finding the path upwards. Leaning into discomfort. All of these clichés are also true. The source of growth is always from our mistakes. Very little growth comes from winning. Embracing our mistakes and failures is vital. Being vulnerable is courageous, and courage is contagious. By sharing our face-down moments, we are cultivating welfare horse sports mindsets in others.

- **Conflict**: Erodes relationships and with the source of conflict, there is an absence of perspective-taking. We can expect to observe that in a psychologically safe CoP, tensions with friction are an important driver for creativity and shaping new ideas.

 REFLEXIVE PRAXIS

Reflect on your experiences that made an equestrian community (yards, clubs, FB groups) psychologically safe. What made another yard/club, FB group unsafe (not speak up, experiment, collaborate, and reflect)? How has status, knowledge (different ways of knowing), and conflict impacted your experience of the equestrian communities you belong to?

BUILDING COMMUNITIES OF PRACTICE

TRUST

Nothing is more valuable in an infodemic era than *trust*. Welfare Horse Sports is a mindset that actively earns trust through small actions to leave horses better, *building* trust every day. Building trust may span different contexts; trust in you (individual), your equestrian community (CoP), and universally (equine sector) share the same observables—making small deposits (not withdrawals) that leave horses better, *because* of our interaction.

Tobias Lutke, founder and CEO of Shopify, likens trust to a battery. When you enter a new equestrian community, your trust may be at 50%. While research shows that when we encounter co-operative behaviour, we experience sensations we call trust. And like a half-charged phone that won't last all day, you have to plug it in. Every interaction you have with people in your community either charges (deposits) or drains (withdraws) the trust battery (see Figure 4.8). Every action you take is an opportunity to build trust or diminish it.

Building trust is made up of four observable behaviours: *reliability*, *accountability*, *integrity*, and *generosity*.

Figure 4.8 The Trust Battery

If our communities align with one or more of our values, trust building starts. Start by showing yourself, community, the universe, who you are, aligning who you claim (say) to be to your actions. When we say we love horses, followed by a small deposit (provide opportunity for a positive experience for the horse), the deposit doesn't just improve the experience for the horse, the experience ripples to equestrians. Serving horses is how we meet our own needs. Helping horses have positive experiences provides equestrians with opportunities to have positive thoughts, and despite flaws being the nature of being human, we mostly don't want to view our flaws, mistakes, and failures. So we resist seeing the gaps between who we say we are and who we want to become for our horses. A welfare horse sports mindset is not about having flaws etc equestrian is not about *not* having flaws, but about trying to become the best version we can, knowing we will never eliminate our flaws. A welfare horse sports mindset is knowing we are all flawed *and* doing constant, endless work in progress.

When we disconnect from who we say we are, what we value (being horse lovers), and we follow up with defensive, blaming, and deflecting behaviours, the disconnect is visible to everyone, rapidly diminishing trust. Doing small deposits for positive change for horses is doing the observable of a welfare horse sports mindset.

If you say you are going to do something for your community, do it. If you need to pivot, explain why. When I make mistakes, being accountable does the opposite of what you think will happen. Owning our flaws and our vulnerabilities makes equestrians *more* trustworthy, not less. Blaming is how we miss opportunities for empathy, spending time making it about whose fault it was. Help your community to be 'able' to hold themselves accountable.

SMALL DEPOSITS

A really effective small deposit to build community and leave horses better is changing your thoughts. Researchers, until recently did not know if negative thoughts triggered negative feelings and if these negative thoughts originated in early childhood. Scientists had come up with different theories, with data conflicting with the different theories. Some findings show that a major component of our thoughts is genetic. Scientists also do not know this for certain. However, the research does indicate that you can change the way you think and feel without having to impact the lived experience of horses.

To build CoPs that leave horses better, it is useful to understand how our thoughts generate our feelings. Using non-recursive modelling (measures causal loops such as what came first, the chicken or the egg?), studies show a causal effect of negative thoughts on negative feelings and positive thoughts on positive feelings. If you have a negative thought and you believe it, you're going to instantly feel lousy. Epictetus claimed thoughts impact feelings 2000 years ago, but studies today have measured the magnitude of thoughts impacting feelings. The thought of being injured by a horse impacts feeling fearful of horses. Feeling fearful (originating from the negative thought) motivates the decision (despite knowing a practice is unethical, such as whipping) and/or equipment (tightened nosebands, relentless rein pressure). The studies also measured the opposite direction, the causal effect of negative feelings on negative thinking. There is an effect, but it is weak, whereas negative thoughts have causal effects on negative feelings. This is a significant finding for horse welfare, understanding resistance to change and why this question—'If you knew better, would you do better?'—matters to positive change for horses.

The process of providing horses with positive experiences *earns* public trust. Positive experiences for horses are the light that shines in very dark corners of today's modern horse training. If the 20th century was the age of the 'deluminator' of horse training (a magical device taken from the Harry Potter books/films); earning public trust is 'illumination for the horse.' We are all responsible for turning the 'light on' (see prediction errors and positive experiences for horses) by taking actions that create a chemical messenger (neurotransmitter in the brain), dopamine, A 'feel-good' signal from an unexpected reward. If the horse is not expecting a scratch at the wither for standing during the mounting process, a better-than-expected experience, dopamine gets released. This surge matters for two reasons; enhancing pleasure *due to the surprise* (consequence of rider mounting is the experience of a scratch—a species-specific need—or a nut) and your better-than-expected scratch/nut encourages learning. Associating the pleasant surprise with standing for the rider to mount makes it more likely to repeat the actions resulting in the reward (scratches or nut). When something positive for the horse happens unexpectedly, the brain's reward system kicks in; dopamine, the brain chemical that makes you feel great and helps horses learn from the experience. Through enriched experiences, positive prediction errors (better than expected) enhance learning whilst also providing evidence of the third central tenet of the Five Domains—provide horses with positive experiences. We cannot infer a good life for a horse when prediction errors (difference between what the horse expected and what actually happened) are negative (e.g., an absence of release of pressure for a behavioural response cued by the rider). We can evidence positive opportunities provided by giving horses better-than-expected outcomes (dopamine release) in every interaction: leave horses better. Positive prediction errors illuminate a path for equestrians to progress their horse towards living a good life (Five Domains and First Principles).

CASCADES OF CHANGE

Researchers show that when it comes to adopting new knowledge, which involves updating our identity, changing an attitude we are currently highly motivated to hold (because it would affect our beliefs in ways that would affect our worldview), or risking our reputation and/or status, change does not go viral in the same way as spreading a meme, TikTok, or YouTube video. We don't need hyperconnected individuals, equestrians who are very good persuaders, or even skilled communicators to spread knowledge (education). According to the research in the field of transformational change, we need lots of leave horses better (LHB) CoPs in order for any equestrian to become the 'spark' that leads to a 'welfare wildfire.' Building CoPs means we get to create lots of 'sparks' in lots of different CoPs. Researchers did not know this until relatively recently. What we now know is that if we want to create change in the world, effectiveness is determined by how well we *anticipate resistance*. In building LHB CoPs, we need to communicate effectively internally and externally across different media platforms strategies to avoid resistance, strategies to overcome resistance when it appears, and what we do when people who don't want the change for horses we want (welfare horse sports) organise in response to growing successful LHB CoPs. To understand the strategy I have adopted to help organisations leave horses better, we need to look at the Greg Satell work on 'cascades.'

Cascades describes what happens when a small set of changes leads to a sequence of change with significant and larger consequences. A psychological network effect involves noticing how others in our CoP are behaving and then choosing to apply learning theory, or not. It involves deciding to join the change that is happening (see Dancing Man Reflexive Praxis). Cascades are founded on three scientific principles:

- *Threshold Model of Conformity.* We all have different thresholds for taking up new knowledge. We all have different thresholds of resistance. Our thresholds for resistance, that is, resistance to conform, is referring to our thresholds of conformity. We conform to other human beings. Looking around our equestrian communities, we tend to follow suit, and the majorities have influence over our behaviour.
- *The Threshold Model of Collective Behaviour.* We are a social species and look around for majorities, then follow suit. The research shows that majorities have influence over our behaviour.
- *The Strength of Weak Ties.* Casual acquaintances have an impact on social networks, on the flow of information, on the flow of change, and behaviour. Horse sports are made up of lots of clusters that know each other well (disciplines) but don't interact much with people outside those disciplines (clusters). Weak ties form *bridges* between clusters of strong ties. The weak tie equestrian between two CoPs becomes the conduit by which information on how to leave horses better, will flow from one CoP to the other.

When we have weak ties between two clusters (disciplines/CoP), if one cluster saturates, it can cause the other cluster that doesn't have the right mix of thresholds to conform and also saturate. Saturation is when everyone in the community follows. Everyone conforms, or almost everyone. That's a saturation within a cluster (CoP). According to the research, all clusters influence each other via weak ties. In LHB clusters (CoPs) of equestrians with strong ties, some of whom are connected to other clusters (CoPs) via weak ties, form what they call a percolating cluster (CoP). Demonstrated by both mathematical models and real-world examinations, mass collective psychology in social change eventually spreads rapidly across entire communities. In equestrianism, the 3 Fs (friends, forage, and freedom) spread rapidly via percolating clusters. Groups of equestrians who were strongly connected *and* weakly connected to other groups in a way that allowed the 3 F's to leap from community to community (discipline to discipline) as thresholds were met and those communities saturated, resulted in clusters of clusters and clusters of superclusters. The 3 F's went 'despacito,' scaling the understanding of friends, forage, and freedom into a massive equestrian-wide status quo-shattering adoption.

Then the new normal levels off until the next cascade comes along. Five Domains (2020) and First Principles (ISES, FEI) cascade via weak connectors in equestrian communities. As equestrians with different interests (disciplines) and goals (ratios of *doing* welfare *and* performance) join and leave equestrian communities, going in and out of contact with the Five Domains (2020) and First Principles (ISES), widespread, large-scale change for horses starts and levels off *inside* CoPs, cascading from one CoP to another until we experience communities percolating horse welfare outcomes (Five Domains, First Principles). Working with the ABRS+ (LHB CoP), I've contributed a series of webinars (on ABRS+ YouTube channel) and facilitated accessibility (discounts for ABRS+ members) to the University of New England's Five Domains online 25-hour learning course, earning CPD points for The British Horse Society Accredited Professionals and Pony Club UK. Clusters of clusters of Leave Horses Better CoP's achieve transformational change for horses via cascades of change. What is the observable in a LHB cascade? Massive equestrian-wide status quo-shattering adoption of First Principles, all starting from weak ties going in and out of LHB communities.

FALSE POSITIVE

Identified by leaders in change movements, the most important area to understand to facilitate positive change for horses is the opposite of what we think it is: a false positive. We think change for horse welfare will happen when there is both more and better *education*. This is only partly true when it comes to actually enhancing horse welfare. Knowing better is not the same as doing better. The idea that equestrians just need better knowledge (facts) to change hearts and minds is incorrect. This is because facts that seem incredibly compelling, the ones that physiologically (releasing rein pressure for deceleration, replacing a pat with a scratch) leave horses better, do not generate change. One reason is that facts do not generate the same emotional reaction in those with a high propensity to *resist* change ('this is how we have always done it' mindset). If you have done the work (unlearning, learning in Chapter 2), your welfare horse sports mindset makes it seem true to you that all we need is more/better education. Researchers found that education (learning new knowledge) can generate even *more* resistance among those already with a high propensity for resistance to change. As an educator for over 20 years (in the science of how horses learn), I have contributed in both small and large ways to translating and disseminating learning theory to equestrians. The truth is, I facilitated both less *and* more resistance to new knowledge. The resistance usually comes back in the form of mental gymnastics, motivated reasoning, and rhetorical strategies that all humans employ in an effort to avoid reaching a conclusion (we now know better for horses, let us do better) or to pursue reaching a different, more favoured conclusion for that individual (confirmation bias). This phenomenon is called the Information Deficit Model and explains why 'education' as *the* solution is not effective on its own at enhancing horse welfare. The lack of mind and heart change for horses in the world is not due to what we think—deficits in our knowledge—or the false negative that there is limited access to knowledge (infodemic era). I have experimented with different ways to communicate the science of how horses learn and landed on the BRAVE acronym. It takes courage to brave taking up new knowledge, doing equitation differently from what we have always done. Hence, I totally understand we all have different degrees of propensity to resist change. The courage it takes to stand alone in the 'wilderness' and enhance the lived experience of horses has been and is now 'home.' I live out there. During these years, I have researched interventions, resources, and communities (in person and online), finding out what actually lowers equestrians' resistance change to apply learning theory and leave horses better.

A resistance to change strategy is one that provides a few small effects which, when added together, lower the propensity to resist change, leaving horses better. A welfare horse sports resistance to change strategy: BRAVE

- *B: behaviour modified*: Demonstrate behavioural modification of the horse. This specifically means ensuring the trainer *feels* the behavioural change of the horse. Two indicators of changed behaviour are the horse responding to your training and feeling *lightness* and responsivity—both trained via *reinforcement* (removal and addition), *repetition* (sets of three, with repetitions in a set of last three improving), and *recency* of training. I call this strategy 'doing what we say we do' as equitation science consultants, behaviourists, trainers, and coaches.
- *R: role model*: Be the change you wish to see horses experience. In all my interactions, I show up providing horses with predicability (light aid), controllability (release of pressure), and self-carriage (trained, not held response to rider/trainer). We don't need to educate *at* others, because we know it does not lower but increases resistance to updating prior knowledge.
- *A: agency*: Acceptance means everyone is on their own equestrian journey. Meeting people where they are and releasing the need to control others can leave horses better. *Doing* modelling is creating a community of choice (agency) as a strategy. A mega-study into gym retention offered two framings (gaining or losing points on non-gym attendance days), with the freedom to choose the framing being the intervention that

improved gym retention. Letting people choose their preferred framing results in human behavioural change (people sticking to the gym).
- *V: values* activity: Identify and synthesise two *core values* through the Five Domains (2020) and First Principles (ISES). Seeing where our values are met in both welfare frameworks lowers resistance to updating what we know.
- *E: empower*: Empower equestrians with opportunities for positive experiences (additional reinforcement in the form of praise from community manager/leader/coach/peer)—be specific. Share with others the specificity of what was done well, leaving horses better. The application of learning theory to people also lowers resistance to change. Positive reinforcement is not just for horses!

Equestrians experiencing a behavioural change in their horses through the correct application of removal and additional reinforcement significantly lowers their resistance to adopting new knowledge. This aligns with the first action in the LHB strategy to lower resistance, which involves feeling and seeing the horse's modified behaviour. The LHB CoP strategy aims to facilitate this reduction in resistance by connecting the organisation's culture (emphasising learning, contemporary practices, heritage understanding, innovation, inclusivity, collaboration, agility, quality, and safety) with the process of lowering resistance. Within a learning culture, individuals who seek new knowledge and learn to apply learning theory, including the use of removal and additional reinforcement, are often those who directly handle the horses, as the organisation prioritises the horse's lived experience. For staff with a higher propensity for resistance to change, a supportive role can be valuable; they can contribute by enhancing the environment to improve the horse's lived experiences. It is important to recognise that resistance to change is a normal human response, and having a strategy within the community to support all staff and empower them to choose to lower their resistance is crucial.

Instead of directly educating those who show resistance, advocate for initiating conversations through open-ended questions. This approach is in line with the principles of MI, which emphasises asking open-ended questions to encourage reflection and dialogue, ultimately making individuals more receptive to new knowledge and ideas. The goal is to understand their perspective rather than imposing a viewpoint. Understanding the underlying reasons for resistance is indeed beneficial. Resistance to change can stem from various factors, including a threat to one's social identity, fear of being judged, or discomfort with ambiguity. Furthermore, resistance can be linked to a lack of introspection. Asking reflexive questions like, 'How does it feel to resist this change?' can prompt self-reflection. This aligns with doing reflexive praxis; taking time to consider one's own beliefs and motivations. This type of questioning can also contribute to developing intellectual humility, which involves recognising the limitations of one's understanding and being open to new knowledge. Cultivating intellectual humility is crucial for a welfare-centred approach to equestrianism, and the BRAVE strategy, discussed previously, also emphasises acceptance that everyone is on their own equestrian journey with meeting people where they are. This aligns with the idea of engaging those with lower resistance first and using open-ended questions to understand different perspectives. The BRAVE strategy fosters change within equestrian communities by focusing on influence rather than control. By starting with those more open to change and using reflective questioning, an environment that encourages voluntary adoption of welfare-enhancing practices cascades a welfare horse sports mindset through CoPs.

Acknowledging the potential for seemingly contradictory viewpoints is crucial for progress in equestrian welfare. For instance, while an equestrian might feel comfortable adhering to familiar practices, they can simultaneously understand that certain adjustments, such as loosening a horse's noseband, are fundamental to the animal's well-being. Given that resistance to change is a natural human response, encountering pushback is to be expected. If a suggestion to improve horse welfare, like loosening a noseband, is met with defensiveness such as, 'Oh, you're one of those coaches,' a helpful approach involves responding with a reflexive question like, 'Yes, I am. May I ask, why wouldn't you be?' Such questions serve as a starting point for the ethical examination of established practices, encouraging self-reflection which can foster intellectual humility. By employing reflexive questioning, particularly with equestrians who demonstrate initial receptiveness, we can guide individuals within our CoPs towards the ethical understanding that inflicting pain on horses for the sake of human ego is unjustifiable. This aligns with the principles of a welfare horse sports mindset which prioritises the horse's lived experience and the application of evidence-based knowledge like the First Principles (ISES) and the Five Domains Welfare Framework.

Acknowledging that resistance to change is a normal human response is fundamental to understanding dynamics within LHB CoPs. Within these communities, individuals exhibit a spectrum of resistance levels, ranging from low to high, when faced with new knowledge and practices. Interestingly, these differences in the propensity to resist evidence-based knowledge can be a catalyst for significant and positive shifts in equestrian practices. When a community embraces a lower threshold for resistance, it often correlates with a move towards more ethical treatment of horses. This transition is further facilitated by cultivating a culture grounded in intellectual humility and introspection, encouraging equestrians to prioritise the horse's lived experience—as guided by the First Principles (ISES) of training and the Five Domains Welfare Framework for management—over personal ego. By centring these evidence-based frameworks, LHB CoPs have the potential to act as 'equestrian "sparks",' igniting and cascading positive change for horse welfare throughout the broader equestrian culture. This aligns with the concept of change spreading through CoPs and the influence of those who readily adopt new, welfare-focused approaches.

 ## REFLEXIVE PRAXIS

- Know and share inside CoPs your two core values (Complete Values Activity on P117)
- Reframe new knowledge as a gain, instead of what we think we will lose by acquiring new knowledge

- Practice synthesising your core values (see above) through each of the ten First Principles. Explore the benefits of the First Principles for the horse (universally and specifically for your horse), the LHB community, and yourself
- Reflect (introspection) on *why* you thought you might lose; money, identity/status, belonging, or other by applying First Principles (ISES)

BECOMING YOUR HORSE'S ADVOCATE

Becoming your horse's advocate is a central theme in cultivating a welfare horse sports mindset, which this book posits as the very heartbeat of ethical equitation. Throughout our reflections in these pages, the consistent truth has been that welfare is not a fleeting trend or a mere checklist item. Instead, it demands integrity, intellectual humility, and, most critically, responsibility—our ability to respond thoughtfully to how horses feel about our interactions. While horses lack a direct voice in policy discussions or public forums, we, as equestrians, possess that voice. Therefore, our words, our actions, and the welfare horse sports mindset we adopt have a profound ability to shape positive change for horses and the future of equestrian sport.

Crucially, genuine progress in horse welfare doesn't occur by simply adding welfare considerations to existing equestrian practices. Instead, it requires a fundamental shift where welfare becomes the very lens through which we interpret, act upon, and evolve all aspects of our engagement with horses. This transformative process is nurtured through the active practices of curiosity, unlearning outdated beliefs, and reflexive praxis, enabling us to prioritise the horse's well-being not just in our words, but through consistent and observable actions.

Reflecting on the insights throughout the book, it is now time to introduce the latest recommendations from leading global equine welfare scientists, presented to the FEI on 9 April 2025 (see Presentation to FEI, April 2025: *https://www.youtube.com/watch?v=W6OgVEC4i28*).

It marks a crucial step towards tangible improvements in horse welfare. These recommendations are described as 'welfare stepping-stones,' intended to inspire equestrians to translate the book's principles into concrete actions that lead to consistently leaving horses better in every interaction. This focus on actionable and meaningful ways to enhance welfare outcomes for all horses is a central theme of the book. The book serves as a blueprint for equestrians curious or currently undertaking shifts in mindset and skills to future-proof horse sports. It aims to move away from traditions not based on evidence and towards a future built on trust earned through consistent, visible welfare improvements. The FEI's past actions, such as unlearning the 'ancestor instinct' and publishing a Welfare Strategy based on the EEWC recommendations in July 2024, indicate a potential receptiveness to these new scientific recommendations. The book itself highlights the growing pressure on horse sports to demonstrate welfare and the need for equestrians to update their skills and mindset to enhance horse welfare outcomes.

The seven recommendations are presented as moral anchors to future-proof sustainable horse sports. They serve as a test of the equestrian community's commitment to prioritise horse welfare beyond rhetoric and translate it into expected practices. This book encourages equestrians to recognise that everyone is responsible for the future of equestrian sports and that optimising and prioritising equine welfare must be visible. Our overstory 'leave horses better' is a call for positive change to both the physical and mental state of sport horses throughout their lives. It involves identifying negative experiences, resolving them, and providing opportunities for positive experiences, ultimately aiming for horses to feel safe, competent, and resilient. This aligns with the Five Domains (2020), a keystone for responsible horse sports.

Real change emerges when welfare becomes the lens through which equestrians interpret, act, and evolve everything done with horses, driven by curiosity, unlearning, and reflexive praxis. The seven recommendations offer a tangible pathway for equestrians, within our own circles of influence, to contribute to a global equestrian community that truly leaves horses better.

1. Accept that any gear that applies pressure is a significant welfare risk
2. Revise judging criteria to reward (not penalise) riders for visible releases of rein tension
3. Recognise pressure-induced tissue compromise, hyperflexion, and signs of fear and pain as evidence of welfare compromise
4. Respect horses' need to perform comfort behaviours (including, but not limited to, lip-licking, coughing, swallowing, yawning)
5. Prevent horse guardians from concealing the horse's attempts to resolve negative experiences (e.g., mouth opening, tongue movement, tail swishing, ear movement)
6. Classify these concealment actions as breaches of 'clean sport' principles
7. Adopt non-invasive tools (e.g., high-definition photography) to identify at-risk horses, and integrate them into judging and stewarding to enable immediate intervention to protect horses, including in 'the field of play'
8. Document pressure-induced lesions and enforce rest until healed

> Do the best you can until you know better. Then, when you know better, do better.
>
> Angelou (2011)

APPENDIX I: LESSON PLANS

LEAVE HORSES BETTER: COMMUNITY OF PRACTICE

Programme of study: Four Training Plans

OBJECTIVE

To empower equestrian centres, riding schools, and grassroots coaches with EBE through the practical application of acceleration (go) and deceleration (stop) responses (in-hand and under-saddle). This training session leverages evidence-based knowledge (First Principles, ISES) *to do* ethical horse training and welfare enhancement.

- *Welfare-centric approach*: This session is grounded in the Five Domains (Mellor et al. 2020), focusing on the horse's *mental* state by providing *predicability* and *controllability* during horse-human interactions. It examines how horses perceive and react to light pressure (predicability) and the necessity for the rider/handler to *release* pressure immediately.

IN THE (PRINCIPLE) SPOTLIGHT

- **Principle 4**: *Regard for Emotional State.* Read the facial and body expression of the horse. Leaving horses better because of our interaction starts by identifying their emotional state. Arousal level with positive/negative experience (valence) provides an objective inference of emotional state.
- **Principle 6**: *Regard for Operant Conditioning (Removal and Additional Reinforcement).* Ensure the correct application of removal (negative) reinforcement of pressure (negative as in mathematical reasoning; the taking away of something aversive) and additional (positive) reinforcement (the adding of an attractive stimulus to reinforce a behaviour). Horses experience predictability of their world through your ability to provide *light* pressures and control what happens to them by your *release* of pressure. Thus, your lightness of pressure and timing of turning off pressure for your horse's mental state is crucial (see Modern Horse Training in Chapter 2).
- **Principle 7**: *Regard for Classical Conditioning (Association Between Cues and Responses).* Understand the importance of classical conditioning, where horses learn to associate a light cue (functions for the horse as a safety signal) with a specific response. Ensure clear, *different* cues for different responses, avoiding confusion (see Modern Horse Training in Chapter 2).
- **Principle 8**: *Correct Use of Shaping (Gradual Development of Complex Behaviours).* Employ shaping gradually in the training of learnt responses towards the desired behaviour by reinforcing *incremental* steps. This allows for more effective and ethical (placing mental state at the centre) training of complex behaviours.
- **Principle 9**: *Correct Use of Signals or Cues (Clarity and Consistency).* Signals/cues/aids must be clear, unambiguous, different for different responses, and be consistent from one trainer to the next. Differing, clashing, or late signals can, at worst, confuse horses, resulting in conflicting behaviour (spooking, bucking, bolting, rearing, and biting). At best, this leads to slow learning.
- **Principle 10**: *Regard for Self-Carriage (Maintaining A Behaviour without Constant Rider Cues).* Train horses to maintain gait,

speed, or posture until cued otherwise, *evidencing* self-carriage. Horses should be trained to learn to sustain behaviour in-hand and under-saddle, until signalled otherwise.

LEARNING OBJECTIVES

- *Understand where each individual horse responds to physical touch*: Explore the area of the base of the wither (where horses groom each other) or other sites of the body that recruit the parasympathetic (relaxation) nervous system as evidenced by increases in eye blink rate/full blinks.
- *Assess the motivation of the horse*: Gauge your horse's motivation to accelerate (go) and decelerate (stop) on the trainer's cue, in-hand and under-saddle; walk, trot, and canter. What is in it for your horse is the *release* of pressure *for* the propulsion or retraction of the legs. If you have decided your horse is unmotivated to go and/or stop, pause and *do* reflexive praxis. Be introspective about your timing of release (what makes their world controllable) and your clarity of your cue (site and lightness).
- *Provide predictability and controllability for your horse* (optimal mental state).
- *Train self-carriage*: Teach horses to maintain gait (keep going) until a signal is given to change, ensuring the absence of constant trainer cueing.

LESSON STRUCTURE

Preparation and Warm-Up

- Use a taper gauge (ISES) to measure the space between nasal plane and noseband. Change equipment/noseband fit for horse comfort

Find their 'sweet spot:'

- Introduce a 'bridge' signal by making a 'cluck' sound before
- Scratching horse at base of wither/either side of the neck
- Condition the bridge by 1. Cluck 2. Scratch 3. Repeat. This is a positive secondary reinforcer

TRAINING THE 'GO' AND 'STOP' CUES

In-Hand Training

- Lead the horse between two markers, assessing lightness (the horse's response to minimal pressure) and responsiveness (how quickly and willingly the horse accelerates or decelerates)
- Acceleration ('go'): Use clear, consistent cues and release this pressure as soon as the horse accelerates
- Deceleration ('stop'): Apply light pressure to stop and slow, releasing pressure the moment the horse stops. Reinforce lightness and responsivity (to get more of both light and responsiveness) with the positive secondary reinforcer (cluck-scratch)
- Test the stop response with a clear two steps back; step back for two steps. The same group of muscles is recruited to slow and step back (pectorals and deltoids)

Under-Saddle Training

- Immobility at the mounting block: Any loss of immobility is an absence of self-carriage in halt. Train 'park' in the absence of self-carriage
- Mount: walking off to the stimulus of the rider mounting (reins picked up, foot in stirrup, rider on) is an absence of self-carriage in halt at the mounting block
- Cue acceleration and deceleration; adults: eight-step exercise. Children: witches' hat (six steps)
- Apply repetition and reinforcement to shape lightness and responsiveness

Timing and Release

- Optimise all horse training by mastering timing of pressure and release. This provides the horse with predictability (clear, consistent, LIGHT cues) and controllability (knowing the response to make/avoid pressure stop)
- Practice applying optimal timing with different horses responsivity; train stop and go with different horses (if a horse has a stop deficit, train a horse with a go deficit) and experience horses with different levels of responsivity (sensitivity to pressure)

How to Leave Horses Better

- Enhance predictability and controllability: Horses should experience clear, different (for different responses), and consistently positioned cues. This reduces confusion and therefore conflict behaviour.
- Reinforce desirable behaviour with additional reinforcement as valued by the horse, not the trainer (scratch or nut).
- Assess and respond: Continuously observe the behaviour of the legs/body/face for arousal and valence (emotional/affect state) to witness the lived experience of the horse which may vary between poor, neutral, and good, fluctuating. Learn to read the variation of mental state between individual horses and build *progress* over perfection, so, over time, each small deposit (leave horses better) progresses the individual horse towards competency, resiliency, and opportunities for positive experiences.
- Self-carriage (absence of force) is not just nice in training, it is an equine ethical mandate. It provides the opportunity for horses to experience controllability of their world (see a mentally more relaxed horse), enhancing their mental state and physically (a more relaxed horse) when observed, the eye identifies and categorises physical and mental relaxation as a 'calm' horse.

THE BIG PICTURE

Every session should aim to not only improve the horse's progression but also leave them mentally and physically better. By understanding and applying EBE, we ensure all horses experience a good life.

REMEMBER

The standard you walk past is the standard you accept.

General David Morrison

GENERAL DAVID MORRISON

By holding ourselves to high standards in horse welfare during training, we contribute to a movement where every horse is left better than we found them.

LEAVE HORSES BETTER

Program of study: Training Plan 2

Foreleg focus: Precision training DIRECT TURNS

OBJECTIVE

To empower grassroots coaches, equestrian centres, riding schools, and livery yards with the knowledge and practical skills (mental and physical) to improve horse welfare by teaching horses to adduct/open forelegs (direct turns) in-hand and under-saddle. This lesson leverages EBE for optimal ethical horse training and welfare enhancement.

WELFARE-CENTRIC APPROACH: KEY CONCEPTS

This session is grounded in the Five Domains (Mellor et al. 2020), focusing on the horse's mental and physical state to promote overall well-being. It examines how horses perceive and react to light pressure (predictability) and the necessity for the rider/handler to *release* pressure immediately.

IN THE (PRINCIPLE) SPOTLIGHT

Principles: 4, 6–10 to underpin this lesson

- **Principle 4**: *Regard for Emotional State.* Read the facial and body expression of the horse. Leaving horses better because of our interaction starts by identifying their emotional state. Arousal level with positive/negative experience (valence) provides an objective inference of emotional state.
- **Principle 6**: *Regard for Operant Conditioning (Use of Positive and Negative Reinforcement).* Ensure the correct application of negative (removal of pressure) or removal reinforcement and positive (additional) reinforcement in training. Horses learn through reinforcement, so your timing and controllability of pressures for horses, are crucial.
- **Principle 7**: *Regard for Classical Conditioning (Association Between Cues and Responses).* Understand the importance of classical

conditioning, where horses learn to associate a light cue (safety signal) with a specific response. Ensure clear, different cues for different responses, avoiding confusion.
- **Principle 8**: *Correct Use of Shaping (Gradual Development of Complex Behaviours)*. Employ shaping to gradually train learnt responses towards the desired behaviour by reinforcing incremental steps. This allows for more effective and ethical training of complex behaviours.
- **Principle 9**: *Correct Use of Signals or Cues (Clarity and Consistency)*. Signals or cues must be clear, unambiguous, different for different responses, and be consistent from one trainer to the next. Differing, clashing, or late signals can confuse horses, at best slowing learning, at worst causing confusion or conflict behaviour; spooking, bucking, bolting, rearing, and biting.
- **Principle 10**: *Regard for Self-Carriage (Maintaining A Behaviour without Constant Rider Cues)*. Train horses to maintain gait, speed, or posture until cued otherwise, evidencing self-carriage. Horses should be trained to learn to sustain behaviour in-hand and under-saddle, until signalled otherwise.

LEARNING OBJECTIVES

- *Understand where each individual horse responds to physical touch*: Explore the area of the base of the wither (where horses groom each other) or other sites of the body that recruit the parasympathetic (relaxation) nervous system as evidenced by increases in eye blink rate/full blinks.
- *Assess the motivation of the horse*: Gauge your horse's motivation to open a foreleg on cue, in-hand and under-saddle; halt, walk, and trot.
- *Provide horse with predictability and controllability*: Ensure your horse receives clear and consistent cues, improving timing (start of swing phase of the forelimb opening/closing) and release of pressure to avoid confusion.
- *Train self-carriage*: Teach horses to maintain direction (keep going) until a signal is given to change direction and/or speed, ensuring the absence of constant cueing.

LESSON STRUCTURE

Preparation and Warm-Up

- Use the ISES taper gauge (optimal space, two fingers) to measure the space between nasal plane and noseband. Change equipment/noseband fit for horse comfort.

Find their 'sweet spot':

- Introduce a 'bridge' signal by making a 'cluck' sound before
- Scratching horse at base of wither/either side of the neck
- Condition the bridge by 1. Cluck 2. Scratch 3. Repeat. This is a positive secondary reinforcer.

Introduce the training device aka 'long whip':

- Introduce training device to horse before session
- Stroke training device all over body (neck/rump/stomach). This is a different feel to the horse than pressure felt from light, fast taps. Use long strokes.
- In halt, with the leg forwards, place the training device on the front of the cannon bone (novel area) and practise light tapping with the device
- As soon as the limb being tapped steps back, STOP tapping
- Repeat. You are assessing your horse's aptitude for removing pressure.
- If horse walks forwards (into taps), DO NOT STOP TAPPING. Whatever the horse does just before you RELEASE pressure, that is, stop tapping, you have reinforced the behaviour before you stopped, that is, walking forwards is how to turn off the taps. This is an opposing behaviour to the one you were training—stepping back to taps on the front of the leg.

TRAINING DIRECT TURNS

In-Hand

- Lead the horse between two markers, assessing lightness (the horse's response to minimal pressure) and responsiveness (how quickly and willingly the horse accelerates or decelerates)

- Decelerate the tempo
- Direct turn (abduct right front leg): Use clear rein/rope cue to the right (direction) followed by pressure, followed by release as soon as horse opens right front leg. Flexing the neck is not a leg response to rein cue. If this happens, take arm/hand further to the right, which reduces neck flexing and increases the size of the right front step
- Bridge (cluck) fore abduction (opening). Reinforce with scratch (sweets pot) reinforcing more of the behaviour (opening) you do want

Under-Saddle

- Explain cue for opening a foreleg (direct turn)
- In halt; cue direct turn by taking rein away from side of neck ('opening a door'). Tap the opposite shoulder to motivate moving away from pressure, that is, adducting/closing foreleg, which then always follows with opening/abduction. STOP tapping when targeted foreleg opens
- In walk, adults ride a 'square': Target cueing at the same time as the swing phase. Riders can look down at the forelimb to align cue with swing phase
- In walk, children ride 'witch's hat'
- Between the corners or marker, repeat go and stop (quicker legs/slower legs) using repetition and reinforcement to shape lightness and responsiveness
- Bridge (cluck) fore abduction (opening). Reinforce with scratch (sweet spot) reinforcing more of the behaviour (opening) you do want
- Adults: Make square bigger—trot
- Children: Make witch's hat bigger—trot
- Progression exercise: Wiggly lines—direct turns

Timing and Release

- Optimise all horse training by mastering your timing of pressure and release. This provides the horse with predictability (clear, consistent, LIGHT cues) and controllability (knowing the response to make/avoid pressure stop).

- Practice applying optimal timing with different horses responsivity; direct and indirect turns are different with different horses (access to a range of movement—opening and closing—varies between breeds, age, and disciplines) and therefore horses will vary in competence, whilst still able to offer improving responsivity (sensitivity to pressure) during sets of three (inside a set, number will vary with the last three improving).

HOW WE LEAVE HORSES BETTER

- Enhance predictability and controllability: Horses should experience clear, different for different responses, and consistently positioned cues. This reduces confusion and therefore conflict behaviour.
- Reinforce desirable behaviour with additional reinforcement as valued by the horse, not the trainer.
- Assess and respond: Continuously observe the behaviour of the legs/body/face for arousal and valence (affective state) to upstand/witness the lived experience of the horse which may range from poor mental state, to neutral, to good throughout the session.
- Self-carriage is a long-term positive experience for horses mentally (light pressure only applied for a new cue as opposed to relentless cueing) and physically; balance providing overall a more relaxed horse who to trainers looks behaviourally 'calm.'

THE BIG PICTURE

Every session should aim to not only improve the horse's progression but also leave them mentally and physically better. By understanding and applying evidence-based practices, we ensure that horses in riding schools and equestrian centres experience a good life.

LEAVE HORSES BETTER

Programme of study: Training Plan 3

Hindleg focus: Precision training LEG-YIELD

APPENDIX I: LESSON PLANS **163**

OBJECTIVE

To empower grassroots coaches, equestrian centres, riding schools, and livery yards with the knowledge and practical skills (mental and physical) to improve horse welfare by teaching horses to close forelegs (indirect turns) in-hand and under-saddle. This lesson leverages science-based knowledge for optimal ethical horse training and welfare enhancement.

WELFARE-CENTRIC APPROACH: KEY CONCEPTS
FOR GRASSROOTS HORSE WELFARE

This lesson is grounded in the Five Domains (Mellor et al 2020), focusing on the horse's mental and physical state to promote overall well-being. Specifically, we will address how horses perceive and react to training pressures and how these pressures can be applied and released ethically.

THE ISES FIRST PRINCIPLES OF HORSE TRAINING

We focus on Principles 6–10 to underpin this lesson:

- **Principle 6**: *Regard for Operant Conditioning (Use of Positive and Negative Reinforcement).* Ensure the correct application of negative (removal of pressure) or removal reinforcement and positive (additional) reinforcement in training. Horses learn through reinforcement, so your timing and controllability of pressures for horses, are crucial.
- **Principle 7**: *Regard for Classical Conditioning (Association Between Cues and Responses).* Understand the importance of classical conditioning, where horses learn to associate a light cue (safety signal) with a specific response. Ensure clear, different cues for different responses, avoiding confusion.
- **Principle 8**: *Correct Use of Shaping (Gradual Development of Complex Behaviours).* Employ shaping to gradually train learnt responses towards the desired behaviour by reinforcing incremental steps. This allows for more effective and ethical training of complex behaviours.
- **Principle 9**: *Correct Use of Signals or Cues (Clarity and Consistency).* Signals or cues must be clear, unambiguous, different for different

responses, and be consistent from one trainer to the next. Differing, clashing, or late signals can confuse horses, at best slowing learning, at worst causing confusion or conflict behaviour; spooking, bucking, bolting, rearing, and biting.
- **Principle 10**: *Regard for Self-Carriage (Maintaining A Behaviour Without Constant Rider Cues).* Train horses to maintain gait, speed, or posture until cued otherwise, evidencing self-carriage. Horses should be trained to learn to sustain behaviour in-hand and under-saddle, until signalled otherwise.

LEARNING OBJECTIVES

- *Understand where each individual horse responds to physical touch*: Explore the area of the base of the wither (where horses groom each other) or other sites of the body that recruit the parasympathetic (relaxation) nervous system as evidenced by increases in eye blink rate/full blinks.
- *Assess the motivation of the horse*: Gauge your horse's motivation to open a foreleg and close a foreleg, on cue, in-hand and under-saddle; halt, walk, and trot.
- *Provide horse with predictability and controllability*: Ensure your horse receives clear and consistent cues, improving timing (start of swing phase of the forelimb opening/closing) and releasing pressure to avoid confusion.
- *Train self-carriage*: Teach horses to maintain direction (keep going) until a signal is given to change direction and/or speed, ensuring the absence of constant cueing.

LESSON STRUCTURE

Preparation and Warm-Up

- Use the ISES taper gauge (optimal space, two fingers) to measure the space between nasal plane and noseband. Change equipment/noseband fit for horse comfort.

Find their 'sweet spot':

- Introduce a 'bridge' signal by making a 'cluck' sound before
- Scratching horse at base of wither/either side of the neck

- Condition the bridge by 1. Cluck 2. Scratch 3. Repeat. This is a positive secondary reinforcer

CHECK: Learnt response to training device aka 'long whip':

- Assess the horse's responsivity to training device in-hand by facing the tail. As the horse walks, at start of swing phase, tap the shoulder for foreleg to close (adduct). Stop tapping when horse closes/adducts
- Repeat so a light touch of the whip on the site of the shoulder moves that limb to horse's midline

TRAINING LEG-YIELD

In-Hand

- Yield: Set your horse up to SUCCEED by halting with the hindlimb to yield (close/adduct) back, that is, at the start of the swing phase. If the limb you want to adduct is under the horse or next to the opposite limb, it is biomechanically harder to respond to your question; close the leg to the midline.
- Using the training device (long whip), tap above the hock (gaskin) with light and fast taps (no gaps)
- No response? Or responds by walking forwards? Apply rein pressure to stop WHILST keeping fast taps. The horse walks forwards (no cue from you to go), and by him clashing the aids (not you), you must continue hindlimb taps (faster, not harder) until horse trials stepping into the midline (adducting).
- You can help success by flexing the head a little towards you in next repetition. (Once they know the correct response [close leg], stop neck flexion.)
- No response?
- Move training device up the limb, above the stifle to the side of the hindleg—Why? Leg-yield site is identical to under-saddle leg-yield taps, giving horse context-specific cue—it will know the answer to the leg-yield question as it adducts (moves away from taps by adducting limb being tapped).
- As soon as the limb being tapped closes (moves to centre of the horse/midline, adducts), STOP tapping.

- Bridge (cluck) fore adduction (closing). Reinforce with scratch (sweet spot) reinforcing more of the behaviour (closing) that you do want.

Under-Saddle Training

Explain cue for closing a hindleg: 10 cm back ('please'). NOT on. No response? Squeeze lower calf ('do it'). No response? Whip tap site of in-hand (shaped up to side of hindleg) ('I mean it'). Hindleg adducts (moves to midline). STOP (release pressure).

- Yield: HALT: Set horse up to succeed by hindleg at start of swing phase
- Slide rider leg back 10 cm from the side of hindleg to yield (close)
- Release as you feel opposite ribs fill into rider boot OR no response?
- Squeeze lower calf
- Release as you feel opposite ribs fill into rider boot OR no response?
- Tap site of side of hindleg. Fast, NOT hard
- Release as you feel opposite ribs fill into rider boot
- As soon as the limb being tapped closes (moves to centre of the horse/midline, adducts), STOP tapping.

Bridge (cluck) for adduction (closing). Reinforce with scratch (sweet spot), reinforcing more of the behaviour (closing) you do want.

- Progression: Centre line and ¾ line leg-yields
- Advanced progression: Leg-yield all around arena (going into and out of the four corners of the arena)

Timing and Release

- Optimise all horse training by mastering your timing of pressure and release. This provides the horse with predictability (clear, consistent, LIGHT cues) and controllability (knowing the response to make/avoid pressure stop).

- Practice applying optimal timing with different horses responsivity; direct and indirect turns are different with different horses (access to a range of movement—opening and closing—varies between breeds, age, and disciplines) and therefore horses will vary in competence, whilst still being able to offer improving responsivity (sensitivity to pressure) during sets of three (inside a set, number will vary with the last three improving).

HOW WE LEAVE HORSES BETTER

- Enhance predictability and controllability: Horses should experience clear, different for different responses, and consistently positioned cues. This reduces confusion and therefore conflict behaviour.
- Reinforce desirable behaviour with additional reinforcement as valued by the horse, not the trainer
- Assess and respond: Continuously observe the behaviour of the legs/body/face for arousal and valence (affective state) to upstand/witness the lived experience of the horse which may range from poor mental state, to neutral, to good throughout the session.
- Self-carriage is a long-term positive experience for horses mentally (light pressure only applied for a new cue as opposed to relentless cueing) and physically; balance providing overall a more relaxed horse who to trainers looks behaviourally 'calm.'

THE BIG PICTURE

Every session should aim to not only improve the horse's progression but also leave them mentally and physically better. By understanding and applying evidence-based practices, we ensure that horses in riding schools and equestrian centres experience a good life.

APPENDIX II: RESOURCES

RESOURCES AND REFERENCES

American Psychological Association (n.d.) Misinformation and disinformation. Available at: https://www.apa.org/topics/journalism-facts/misinformation-disinformation (Accessed: 8 August 2025).

Angelou, M. (2011, January 1). Quotation from *Oprah's Master Class* [Television broadcast]. Oprah Winfrey Network.

Ashton, L., Brown Watson, J. and Hartmann, E. (2025) 'Bridging theory and practice: How a community of practice can drive cultural change in equitation science and welfare education'. Poster presented at 20th International Society for Equitation Science Conference, Colorado State University, Fort Collins, 15–18 July. In: Proceedings of the 20th International Society for Equitation Science Conference, poster no. 14, p. 77.

Belasco, J. A., & Stayer, R. C. (1994). *Flight of the buffalo: Soaring to excellence, learning to let employees lead.* Grand Central Publishing.

Boyer, E. L. (1990). *Scholarship reconsidered: Priorities of the professoriate.* Carnegie Foundation for the Advancement of Teaching.

Chandler, D.B. (2001) 'Resolving and managing conflicts in academic communities: The emerging role of the "pracademic"', International Journal of Conflict Management, 12(2), pp. 161–172.

Cicero, M. T. (1913). *De officiis* (W. Miller, Trans., Classical Library, Vol. 30). Harvard University Press.

Dutch Animal Protection Act Amendment. (2021). Wijziging van de Wet dieren in verband met de uitvoering van de herziene Europese diergezondheidswetgeving – Gewijzigd amendement van het lid vestering ter Vervanging van dat gedrukt onder nr. 9, Ontvangen 11 mei 2021, Tweede Kamer, vergaderjaar 2020–2021, 35 398, nr. 22.

Eurogroup for Animals. (2024). Good welfare for equids: White paper. Eurogroup for Animals, in collaboration with the Donkey Sanctuary.

Equine Ethics and Wellbeing Commission. (2022). Further supporting information related to Tack and Equipment Early Recommendations: Review of relevant research (August–November 2022) [Online report]. Equine Ethics and Wellbeing Commission.

Facciani, M. (2025). *Misguided: Where Misinformation Starts, How It Spreads, and What to Do about It*. New York: Columbia University Press.

Fiedler, J. M., Ayre, M. L., Rosanowski, S., & Slater, J. D. (2025). Horses are worthy of care: Horse sector participants' attitudes towards animal sentience, welfare, and well-being. *Animal Welfare*, 34, e6, 1–12. https://doi.org/10.1017/awf.2024.69

Fraser, D., Weary, D. M., Pajor, E. A., & Milligan, B. N. (1997). A scientific conception of animal welfare that reflects ethical concerns. *Animal Welfare*, 6(3), 187–205.

Grant, A. (2025) Interview at Brilliant Minds Conference. In conversation with A. S. Instagram Reels, 16 September. Available at: https://www.instagram.com/reel/DOtHCwNkaa6/?igsh=ZXdiM2RxbHIwY3J6 (Accessed: 16 September 2025).

Guerdat, S. (2025). Contribution to panel discussion 'Coaching in the Future' [Unpublished]. The Future of Equestrian Coaching Symposium, Royal Dublin Society, Dublin, 17 April.

International Society for Equitation Science. (2018). Principles of learning theory in equitation. https://www.equitationscience.com/ises-training-principles

Jobs, S. (2023). *Make something wonderful: Steve Jobs in his own words* (L. Berlin, Ed.). Steve Jobs Archive.

Jones, P. & Warren, S. (2016). *Horses hate surprise parties: Equitation science for young riders*. Equitation Science International.

Jones McVey, R. (2023). Human-horse relations and the ethics of knowing. Routledge.

King, M. L. Jr. (1965). How long? Not long, speech delivered at the conclusion of the Selma-to-Montgomery March, Montgomery, Alabama, 25 March. In: Our God is Marching On! [Transcript].

Lesté-Lasserre, C. (2023, December 12). FEI equine ethics and well-being commission delivers final recommendations. *Chronicle of the Horse*.

Littlewood, K. E., Heslop, M. V., & Cobb, M. L. (2023). The agency domain and behavioral interactions: Assessing positive animal welfare using the Five Domains Model. *Frontiers in Veterinary Science*, 10, 1284869. https://doi.org/10.3389/fvets.2023.1284869

Lubelska-Sazanów, M. (2024). Ethics in sports industry: When does sports autonomy become an excuse for animal abuse? *International Journal of Law in Context*, 1–17. https://doi.org/10.1017/S1744552324000193

Maurício, L. S., Leme, D. P., & Hötzel, M. J. (2024). The easiest becomes the rule: Beliefs, knowledge and attitudes of equine practitioners and enthusiasts regarding horse welfare. *Animals*, 14, 1282. https://doi.org/10.3390/ani14091282

McGreevy, P.D., Mellor, D.J., Freire, R., Fenner, K., Merkies, K., Warren-Smith, A., Uldahl, M., Starling, M., Lykins, A., McLean, A., Doherty, O., Bradshaw-Wiley, E., Quinn, R., Wilkins, C.L., Christensen, J.W., Jones, B., Ashton, L., Padalino, B., O' Brien, C., Copelin, C., Brady, C., Henshall C. 2025. COMPASS GUIDELINES for conducting welfare-focused research into behaviour modification of animals. Animals. Under review

McGreevy P., Winther Christensen, J., König von Borstel, U., & McLean, A. (2018). *Equitation science*. Wiley-Blackwell.

McLean, A. N. (2003). *The truth about horses*. Penguin Books.

McLean, A. N. (2024). Modern horse training: Equitation science principles & practice, Volume 2—Principles in practice. Equitation Science International.

McLean, A., McGreevy, P.D. & ISES Council (2018). The new ISES training principles. In: S. McDonnell, B. Padalino & P. Baragli (eds.) Proceedings of the 14th International Conference of the International Society for Equitation Science: Equitation Science 150 years after Caprilli – theory and practice, the full circle, Rome, Italy, September 21–24, 2018. Pisa: Pisa University Press, p. 11. ISBN 978-88-6741-888-6.

McLean, A. N., & McLean, M. (2008) *Academic horse training*. Australian Equine Behaviour Centre.

Mellor, D. J., Beausoleil, N. J., Littlewood, K. E., McLean, A. N., McGreevy, P. D., Jones, B., & Wilkins, C. (2020). The 2020 five domains model: Including human–animal interactions in assessments of animal welfare. *Animals*, 10, 1870. https://doi.org/10.3390/ani10101870

Ödberg, F.O. and Bouissou, M.F. (1999) 'The development of equestrianism from the Baroque period to the present day and its consequences for the welfare of horses', Equine Veterinary Journal, Supplement 28, pp. 26–30.

Pink, D. H. (2022). *The power of regret: How looking backward moves us forward*. Riverhead Books.

Sinek, S. (2010, February 24). Great leaders don't see themselves as great; they see themselves as human. X (formerly Twitter). https://x.com/simonsinek/status/9579173952

The Times. (2024, September 27). Charlotte Dujardin and equine welfare. The Times (UK).

Triple WellBeing® Framework. (n.d.). "You're always practising something. So you're either practising upholding the world as it is, or you're practising shifting into the world as you want it to be." —Adrienne Maree Brown. Thoughtbox Education.

Volpe, M.R. and Chandler, D.B. (2001) 'Resolving and managing conflicts in academic communities: The emerging role of the "pracademic"', International Journal of Conflict Management, 12(2), pp. 161–172.

World Organisation for Animal Health (WOAH) (2024) Animal welfare: a vital asset for a more sustainable world. Vision Paper. Paris: WOAH. Available at: https://www.woah.org/app/uploads/2024/01/en-woah-visionpaper-animalwelfare.pdf (Accessed: 26 September 2025)

Zetterqvist Blokhuis, M. (2021). Teaching horse riding: Is the role of the horse recognized? Society & Animals, pp.1–17. doi:10.1163/15685306-bja10062.

ONLINE RESOURCES

Equitation Science International Diploma

First Principles (ISES)

Coffee With Horse Lovers

- Download any free metronome application from your App Store.
- Tap the button on the app every time the same limb contacts the surface. The metronome supports your coaching of variation of speed—'gears within gears.' It will help you coach speed control following variations of gait tempo. Courtesy of Dr Andrew McLean.
- The following beats per minute (bpm) is your guide to train 'gears within gears' aka tempo control.

- Walk 50–60 bpm—average 55 bpm
- Trot 70–80 bpm—average 65 bpm
- Canter 95–110 bpm—it varies depending on function, that is, dressage, show jumping, cross-country, etc.

INDEX

Abundance
    concept of, 5, 80, 132; *see also* Flourish; Positive affect

Abuse
    concepts of, 50, 98, 109
    cultural and systemic forms of, 50, 97; *see also* Harm; Welfare violations

Academic
    contexts of, 65, 105
    pracademic identity, *see* Praxis–reflexive

Acceleration
    concept of, 6, 13, 69, 74, 77, 138, 153, 155, 156, 160; *see also* Change; Momentum

Acceptance
    as cultural shift, 146; *see also* Adaptation; Trust
    in human–horse relationship, 37, 93
    as learning state, 37, 47, 93, 97, 98, 127, 146

Accommodation
    concepts of, 3, 40, 42; *see also* Assimilation; Learning theory

Accountability
    concepts of, 9, 87, 106, 114, 125, 130, 139; *see also* Ethics; Responsibility

Activism
    concepts of, 108–109; *see also* Advocacy

Activities
    learning and practice activities, 8, 13, 20, 23, 35–36, 65, 108, 118–119, 138, 147, 149; *see also* Coaching; Teaching

Adapt
    concepts of, 3, 40, 89; *see also* Change; Flexibility

Additional reinforcement
    concept of, 69, 74, 76, 78, 147, 154, 157, 158, 162, 163, 167; *see also* Operant conditioning; Positive reinforcement

Adduction
    in biomechanics and movement, 72, 77, 158, 161, 165, 166

Adversity
    concepts of, 79, 123, 125; *see also* Challenge; Stress

Advocacy
concepts of, 69, 106, 107, 134, 136; *see also* Activism; Leadership
Affect
concept of, 5, 17, 54, 60, 66, 77, 92, 122, 130, 142, 157; *see also* Emotion; Feeling
Affective states
concepts of, xv, 26, 162, 167; *see also* Emotion; Five Domains
Agency, *see* Control; Controllability; Empowerment
Aids
agentic behaviour and expression, 48, 132
concept of, 7, 18, 29, 37, 73, 154, 165
training and communication aids, 7, 17, 29, 65–66, 153, 163
Ally
concept of, 4, 28, 94; *see also* Collaboration; Partnerships
Ambiguity
concepts of, 67–68, 148; *see also* Clarity; Uncertainty
Ancestry
concept of, 12, 83; *see also* Heritage; Tradition
Angelou, Maya
quotations and references, 40, 152
Anthropomorphism
concept of, viii, 66, 70–72, 92–93
risks of, 66, 70–72; *see also* Empathy; Human projection
Anxiety
concepts of, 27, 66, 82, 97; *see also* Arousal; Stress
Arousal
concept of, 27, 44, 66, 71, 76, 79, 154, 157–158, 162, 167; *see also* Emotion; Stress
Assimilation
concepts of, 3, 40; *see also* Accommodation; Learning theory
Assumptions
concept of, 5, 24, 50, 53, 108; *see also* Bias; Epistemology
Attachment
concepts of, 53, 70; *see also* Bond; Relationship
Attitudes
concept of, xviii, 1, 3, 5, 8–10, 12, 23, 25, 32, 40, 41, 98, 135; *see also* Beliefs; Culture
Authenticity
concepts of, 112, 119, 128, 132; *see also* Integrity; Trust
Aversive
stimuli and training methods, 17, 49, 71–72, 76–78, 154; *see also* Negative reinforcement; Punishment
Avoidance
behaviour and responses, 4, 8, 13, 18, 20, 71, 79, 85, 98, 154, 159, 161; *see also* Aversive stimuli; Fear
Awareness
concept of, 61, 85, 87, 89, 94, 109; *see also* Consciousness; Reflection

Balance
concepts of, 3, 15–16, 24, 32, 162, 167
psychological balance in horse–human interaction, 15–16;

INDEX

see also Equilibrium; Harmony

Barriers
  concepts of, 22, 102, 128, 129; *see also* Challenges; Obstacles

Behaviour
  belief systems shaping practice, 3, 39–43, 53, 148; *see also* Assumptions; Culture
  concepts of, ix, xiii–xiv, xviii, 1–2, 4–9, 11, 13, 18, 22, 24–25, 28–30, 32–34, 38–50, 53, 56, 59, 67, 70–78, 84–86, 91, 96–103, 108, 111, 116, 122, 135, 137, 142, 146–147, 152, 154, 155, 157, 164, 166
  as indicator of welfare, 32, 55
  learning and behaviour change, 36, 40, 48, 50; *see also* Beliefs; Learning; Training; Welfare

Belonging
  communities of belonging, 59–61, 150; *see also* Community; Connection
  concepts of, 11, 23, 53, 59–61, 88, 97–99, 132, 150

Biases
  concepts of, 5, 11, 89; *see also* Assumptions; Critical thinking

Biomechanics
  application in equitation, 7, 93
  concept of, 4, 7, 93; *see also* Movement; Physiology

Biting
  behavioural expressions of, 154, 159, 164; *see also* Aggression; Avoidance

Blink eye rate
  as welfare indicator, 155, 159, 164; *see also* Behaviour, indicators; Stress

Blueprint
  concept of, 1, 12, 151; *see also* Framework; Model

Bolting
  behavioural expression of, 4, 12, 91, 154, 159, 164; *see also* Fear; Flight response

Bond
  concepts of, 35, 42, 59, 122; *see also* Attachment; Connection

Brain
  concepts of, 3, 11, 28, 36–37, 40, 42, 44, 85, 88, 103, 141
  learning and neurobiology, 36–37, 42, 44, 85; *see also* Cognition; Neuroscience

Brave
  concepts of, 70, 107, 136, 146; *see also* Confidence; Courage

Bridle
  concepts of, 55, 60
  use of, 18, 37, 55, 60; *see also* Abuse; Bit; Brutalization; Harm; Noseband; Tack

Bucking
  behavioural expression of, 4, 8, 91, 154, 159, 164; *see also* Flight response; Stress

Calm
  calm states in learning, 24, 157
  concept of, 24, 157, 162, 167; *see also* Arousal; Relaxation

Capability
  concepts of, 62, 64, 100, 102, 111; *see also* Capacity; Competence

Capacity
  concepts of, 3, 48, 81; *see also* Capability; Resilience

Cascades
  concept of, 9, 96, 142, 143, 148

realist ripples as cascades of change, 96, 143; *see also* Change; Culture

Certainty
concepts of, 10, 23, 41, 51, 54, 68, 110; *see also* Control; Predictability

Challenges
concept of, 4, 9–11, 26, 34, 43, 61, 65, 69, 89, 99, 100, 105–106, 112, 138; *see also* Adversity; Barriers

Change
concepts of, 1–3, 6, 8–11, 14, 19, 24, 27, 32–40, 43–47, 50, 52–53, 58–59, 70, 75–76, 81–83, 89, 91, 93, 96–101, 103–104, 107, 109–111, 121–123, 128, 135, 140–151, 155, 160, 164
cultural and systemic change, 70, 121–123; *see also* Adaptation; Transformation

Checklists
concept of, 17, 19, 76, 115, 150; *see also* Framework

Choice
concepts of, 5, 15, 24, 26, 34, 36, 38, 45, 48, 55, 60, 68, 75, 91, 114, 146; *see also* Agency; Control; Learning

Clarity
concept of, 18, 38, 43, 67, 72, 154–155, 159, 163; *see also* Communication; Transparency

Classical equitation
concepts of, 1, 6, 7, 10, 12, 20, 25, 28, 34, 52, 65, 66, 85–88, 97, 115–116, 146, 150; *see also* Training, Traditional

Cluck
use of, 13, 73–74, 78, 122, 155, 160–161, 164–166

Cluster
concepts of, 143–144

Coaches
as practitioners, 12, 26, 57, 66, 85, 90, 115, 146, 148, 153, 158, 163; *see also* Coaching; Educators

Coaching
approaches to, 6
practice of, 26, 34, 38, 69
roles in equestrian culture, 125
skills of coaches; *see also* Coaches

Coercion
concepts of, 26, 67, 76; *see also* Aversive, stimuli and training methods; Control

Cognition
applied in equitation science, 40, 42, 86, 88, 93
concept of, xiii, 15; *see also* Brain; Learning

Collaboration
concept of, 5
practices of, 65, 86, 131–133, 135, 147; *see also* Communities of practice; Partnerships

Collection
in horse training, 3–4, 40; *see also* Balance; Biomechanics

Comfort
experience of being comfortable, 41, 82, 148
as welfare state, 11, 18, 54, 66, 152, 155, 160, 164; *see also* Welfare

Communication
acts of communicating, 45, 51
concepts of, 18, 38, 43, 69, 72, 87, 99, 102, 104, 119;

see also Cues; Signals; Feedback

Communities of practice (CoP)
  mentions of, 69, 96–98, 105, 133–135, 139–142, 147–148; *see also* Collaboration; Community; Learning

Community
  concepts of, 12, 14, 42, 48, 59–60, 65, 68–69, 84, 86, 96–98, 102, 104–105, 113, 130, 133, 134, 136, 138–140, 144, 149, 150; *see also* Communities of practice; Culture

Complexity
  concepts of, 9, 11, 29, 64, 67, 89, 94, 96, 97, 100, 132, 154, 159, 163; *see also* Systems

Concepts
  concerns as concepts, 58
  general use, 28, 61, 70, 93, 97, 125, 149

Conditioning
  classical and operant, 49, 73–74, 95–96
  in equestrian training, 154, 158–160, 163; *see also* Learning theory; Reinforcement

Conflict
  conflicting perspectives, 4, 141, 154
  conflicts in practice, 20–21, 79, 112, 138, 154, 159, 162, 167
  in horse–human interaction, 4–5, 7–8, 12, 36, 49, 60

Confusion
  causes of, 7, 10, 12, 18, 27, 50, 71, 79, 85
  expressions of (confuse/confused), 154, 157, 162–164, 167

Connection
  acts of connecting (connect/connected/connecting), 111, 127–129, 132
  concept of, 1, 41–42, 48, 52, 70, 97–99, 111, 127–129, 132, 137
  multiple connections, 53, 97–99, 132

Consequences
  of behaviour and training, 33, 79, 143

Consistency
  acting consistently, 154, 159
  applied in equitation practice, 163
  concept of, 154, 159, 163

Contact
  concepts of, 39, 96, 144

Control
  controllability, 34, 49, 64, 146, 153–157, 159–162, 164, 166
  controlling actions, 111
  general concepts, 17, 19, 23, 25, 31, 33, 38, 44, 46–49, 52, 60, 72, 75, 90, 102–103, 111–112, 146, 147, 153

Conversation
  concept of, 14, 45, 47, 51–52, 68, 69, 93, 113, 120
  conversations in practice, 14, 41, 48, 51–52, 84, 104, 113–114, 120, 134, 147

Courage
  courageous actions, 106, 127, 138
  developing in equestrians, 52, 106–107, 109
  expressions of, 119, 120, 127–128, 138, 146

Critical
  concepts of, 25, 47, 61–62, 70–71, 75–76, 86, 89, 92–93, 96, 99, 127, 150

critically applied, 10, 24–25, 70, 89
Cross-country
contexts of, 18, 70
Cues
concept of, 4, 7, 18, 61, 72–74, 95–96, 154, 156, 159, 161–164
cued responses, 154, 156, 162
cueing in equitation, 95–96, 154, 156, 159, 161, 162, 166
Cultivate
acts of cultivating (cultivate/cultivating), 36, 51, 97, 113, 127, 132, 135
Culture
concepts of, 11–12, 26, 39, 43, 50, 58–60, 68, 81, 83, 89, 111, 118, 135, 147, 149
cultural dimensions and contexts, 10, 11, 70, 71, 96; *see also* Community; Tradition
Curiosity
concepts of, xviii, 5–6, 14, 23, 24, 27, 40, 45, 47–53, 68, 78, 81, 90, 99, 106, 119, 136, 150–151
curious expressions, xviii, 5, 24, 40, 48–53, 78, 90, 106, 151
Curriculum
development of, 81, 83, 92

Debunking
examples of, 88, 94
Deceleration
concepts of, 28, 49
in horse training, 145, 153, 156
Deception
concepts of, 85, 87
examples, 93–94

Decision-making
concepts of decision and deciding, 55, 90–91, 120, 130, 141
deciding in practice, 4, 13, 15–16, 26–27, 33, 39, 90, 91, 103
relationship to behaviour change, 55, 141
Detective
as metaphor for reflexive praxis, 75, 89, 103
Development
concepts of, 5–6, 35, 48, 70, 81, 105, 108, 154
developing practices, 108, 111, 113
Discernment
concepts of, xvii–xviii
Discipline
as concept, 41, 48, 112, 143
equestrian disciplines, 12, 64–65, 90, 97, 143–144, 162, 167
Discomfort
in horses and riders, 1, 4, 5, 25, 27, 36, 38, 47, 49, 50, 54, 55, 61, 68, 99, 122, 129, 138, 148
Disconnect
concepts of, 47
in equestrian contexts, 119, 140
Disinformation
concepts of, 57–58
Disruption
concepts of, 107–108
in equestrian contexts, 136
Distress
concepts of, 15, 20, 38
examples, 82, 109
Diversity
concepts of, 5, 9, 22, 32, 60, 61, 65, 89, 104–108, 134
Domain
concept of domains, 5, 29–31, 48, 62, 64, 78, 136

Five Domains, *see* Five Domains
Dominance
  as concept, 50, 66, 116
Dopamine
  in learning and motivation, 28–29, 51, 74, 75, 77, 141
Dressage
  concepts of, 18, 25, 27, 39
  practice and culture, 39, 60, 89, 90

Edmondson, Amy
  psychological safety and learning, 134–135
Education
  concepts of, xii–xiii, 6, 8–9, 32, 61, 70, 81, 83, 90–92, 94, 99, 100, 102, 104, 115, 118, 136, 142, 145
  educating in equestrian practice, 32, 61, 102, 149–150
  educational approaches, 5, 13, 149
Emotion
  concepts and emotional dimensions, xv, xviii, 3, 13, 16–18, 40, 54, 77–79, 122, 124, 127, 132, 137, 145, 154, 157, 158
  emotions in horse training, xv, 66, 79, 92, 115, 122, 123
Empathy
  concepts of, ix, 11, 59, 72
  in equestrian practice, 32, 61, 102–103, 149
Empowerment
  concepts of, xviii, 10, 102, 147, 153, 158, 163
Enclosures
  use and critique of, 8, 32
Enhancement
  of welfare and practice, 19–22, 54, 107, 153, 158, 163
Enrichment
  concepts of, 5, 35, 123, 124
Entertainment
  uses of animals in, 8, 32, 87
Epistemology
  concepts of, 5–6, 26, 92, 137
Equi-centric
  approach and mindset, 8, 27, 57, 62
Ethics
  concepts and debates, xiii, xvi, xvii, 12, 57, 61, 102, 111
  ethics and public trust within Five Domains, 13, 23, 24, 62, 92, 106, 118, 119, 141
Ethology
  references to, 15, 93

Facilitation/Facilitate/Facilitating, 19, 20, 28, 66, 74, 96, 99, 104, 125, 136, 138, 144–145, 147, 149
Failure/Failing/Fails, 68, 77, 100, 112, 138, 140
  concepts of, 81, 111, 138
  examples, 3, 7, 126
False
  concepts of, 10, 56–57, 86, 88, 145
  falsehoods, 10, 57–59
Fear
  concepts of, xvii–xviii, 5, 7, 18, 38, 68, 71, 82, 85, 97–99, 104–105, 107–110, 112, 122, 127, 135, 141, 148
  fearful states, 18, 71, 76, 110, 141
Feedback
  concepts and use of, 22, 60, 73, 86, 121, 134
First Principles (ISES)
  concepts of, xix, 6–7, 13, 24–25, 31, 36, 43, 49–51, 64, 66, 69, 70, 73–76, 78, 79, 81, 83–85, 87, 92, 96–100, 102, 108, 115, 116, 122, 136, 137, 142, 144, 145, 147, 149, 150, 153, 163
Five Domains (Mellor)
  concepts of, xii, xix, 27, 30–32, 40, 48, 61, 62, 64–65, 69, 78, 81, 84, 99, 108, 136,

138, 142, 144, 147, 149, 151, 153, 158, 163

Flourish
  concepts of, 34, 35, 72, 82, 104
  flourishing, 35, 80, 81, 105

Forage, vii, 18–21, 64, 78, 80, 123, 144

Fortitude
  concepts of, 35, 39, 64, 77–78

Framework
  concepts of, xix, 3–5, 12, 15, 18, 32, 34, 40, 49, 56–57, 64, 65, 80, 81, 83, 147, 149
  frameworks in practice, 57, 65, 116
  framing, 8, 29, 38, 77, 135–136, 146–147

Friend(s)
  concepts of friendship, 16, 18–19, 33–35, 37, 39, 44–47, 52–53, 59, 64, 78, 80, 115, 118, 123, 128–130, 144

Gauge
  noseband gauge, use of, 13, 155, 159–160, 164; *see also* Measurement; Noseband

Grassroot
  concepts of, 82, 90, 102, 107, 153, 158, 163; *see also* Community

Grazing
  behaviour and management, 20–21, 123; *see also* Environment; Feeding; Foraging

Grooming
  concepts of, 11, 19–20, 25, 39, 42, 80; *see also* Social behaviour; Touch

Growth
  concepts of, 4, 38–39, 54, 68, 89, 124, 138; *see also* Change; Development

Guardians
  concepts of, 31, 86, 150; *see also* Responsibility; Stewardship

Guard-rails
  concept of, 70, 134; *see also* Boundaries; Safety

Habits
  concepts of, 25, 67, 121; *see also* Behaviour; Routines

Halt
  as ridden transition, 18, 35, 156, 159–161, 164, 166; *see also* Cues; Transitions

Hand
  use in riding and handling, 39, 44, 54, 72–77, 86, 147, 159; *see also* Aids; Contact

Handler
  role of, 18, 76, 153; *see also* In-hand; Training

Hans, Clever
  references to, viii, 88, 92–96; *see also* Cognition; Communication

Harmony
  concepts of, 20, 81, 105; *see also* Balance; Partnership

Health
  concepts of, 19, 29–31, 34, 42, 84, 100–102, 109, 111, 121, 126, 133–134; *see also* Welfare; Well-being

Healthy
  concepts of, 77, 80–81, 134; *see also* Health; Thriving

Heartland
  concepts of, 2, 27, 41, 48, 50

Herd
  social organisation, 20–21, 102; *see also* Social behaviour

Heritage
  concepts of, xvi–xvii, 57, 60, 89, 147; *see also* Tradition

Human/Humans
concepts of, xvi–xvii, 3, 8–11, 13, 16–18, 23, 26–29, 31, 33–34, 37, 39–41, 48, 55, 57–59, 68, 70, 72–74, 76–79, 81–82, 91–99, 101–104, 110–111, 122, 131–132, 140, 143, 145–148; *see also* Anthropocentrism

Humility
concepts of, 17, 40, 48, 51, 57, 59, 61, 68, 113, 148–150; *see also* Empathy; Reflection

Hyper-flexed
concepts of, 1, 109; *see also* Biomechanics; Posture

Identity/Identities (social/professional), 53, 59–60, 68, 92, 99; *see also* Roles; Ontology
concepts of, 11, 51, 56, 58–60, 64, 68, 70, 85, 88, 97–99, 142, 148, 150

Immobile/Immobility
states of, xiv, 35, 44, 75–76, 156; *see also* Posture; Stillness

Indirect
indirect effects and methods, 162–163, 167

Inference
inference and interpretation, 72–75, 79, 93, 154, 158

Influence/Influences/Influencing
acts of influencing, 44, 52, 96, 115
concepts of influence, 4, 29, 33–34, 39, 44–46, 66, 68–70, 89, 97–98, 111–112, 134, 143, 148–149, 151; *see also* Culture; Impact; Leadership

Infodemic
concepts of, 10, 64, 109, 139, 145; *see also* Communication; Misinformation

In-hand
work and handling, 43, 76–77, 123, 153, 155–156, 158–160, 163–166; *see also* Handler; Groundwork

Innovation/Innovations
innovation in practice, 43, 66, 78, 84, 97, 104, 113, 147; *see also* Change; Ideas

Inspire/Inspired
inspiration and motivated action, 37, 39, 44, 104–105, 138, 151

Interconnectedness
concepts of, 84, 96, 111

International Society for Equitation Science (ISES)
references to, vii, xii, xvii–xix, 7, 25, 31, 51, 54, 62, 64–65, 69–70, 73–74, 76, 78–79, 81, 83–85, 87, 90, 92, 96–100, 102, 108, 115–116, 122, 136–137, 144, 147, 149–150, 153, 155, 160, 163–164; *see also* First Principles (ISES)

Intersectional/Intersectionality
concepts of, xiii, 2, 8, 70, 92

Intrinsic
concepts of, 118

Introspection
reflection and introspection, 2, 6–7, 34, 75–76, 148–150; *see also* Reflection; Reflexive praxis

Journaling
act of journaling, 22, 125
concepts of, xiii, 19, 59, 85, 91, 125; *see also* Reflection; Reflexive practice; Reflexive praxis

Journeys
concept of, v, 6, 10, 44, 50, 99, 108, 113, 119, 125,

130, 137, 147–148; *see also* Learning; Transformation; Reflection

Judge/Judgeded/s/Judging/Judgement
concepts of, xvi, 4, 12, 25–26, 37, 42, 46, 52, 61–61, 68, 71, 77, 85, 88, 103, 126, 148; *see also* Assessment; Competition; Decision-making; Evaluation; Perception

Jump/Jumping
concepts of, 19, 27, 37, 90, 103–104, 113; *see also* Cross-country; Equitation; Training

Justice
concepts of, 33, 62, 110; *see also* Ethics; Fairness; Moral

Kind/Kindness
concepts of, 13, 85, 129; *see also* Compassion; Empathy

Knew/Known/Knows/Knowing
concepts of knowledge and awareness, 4, 5, 10, 16, 24, 28, 33–35, 40, 46, 48, 50, 54, 55, 58, 64, 65, 72, 77, 78, 82, 86, 88, 94, 96, 103, 110, 115, 118, 122, 132, 137, 141, 148, 156, 161, 166; *see also* Epistemology; Reflection; Understanding

Knowledge
concepts and applications of, 3–8, 10–11, 13–17, 23–26, 29, 31, 38–41, 48–57, 59, 64–65, 69–72, 76, 79, 81–83, 87, 92, 96–100, 102, 104, 107–108, 116, 135–138, 142, 145–149, 153, 158, 163

knowledge exchange and sharing, 11, 23–26, 64, 87, 97–99, 147–149; *see also* Evidence; Expertise; Learning

Language
concepts of, 20, 76, 80, 95, 125, 127; *see also* Communication; Terminology

Laws, 32, 80, 88, 127, 140

Layers
concepts of layering, 58, 63–64

Lazy
concepts of, xviii, 8, 37, 71, 103

Lead/Leading/Leads
acts of leading and guidance, xiii–xv, 7, 18, 31, 43, 45, 58, 59, 69–74, 81, 85, 91–93, 97–98, 103; *see also* Guidance

Leaders
concepts of, viii–ix, 32, 68, 81, 91–92, 97–98, 110–114, 120, 135, 145, 147; *see also* Coaching; Influence

Legal/Legally, 8, 9, 32, 62, 76; *see also* Laws
legislation and legal contexts, 8–9, 32, 62, 76; *see also* Governance; Policy

Lens
concepts of, xvi–xvii, 1, 13, 16, 62, 71–72, 96, 133, 149–150; *see also* Perspective; Framework

Lessons
contexts of teaching and learning, ix, 20, 61, 104, 114–115, 153–154, 158, 163; *see also* Education; Coaching

Leverage
concepts of, 153, 158, 163; *see also* Aids; Mechanics

Liberty
  concepts of, 74–75; *see also* Agency; Freedom
Licence, *see* Governance
Lies
  concepts of, 1, 60, 124, 136; *see also* Authenticity; Truth

Magic, 44, 85–86
Magicians, 85–86, 88, 93–94
Mandate
  references to, 125, 136, 157
Mastering
  skills and mastery, 156, 161, 166
Mathematical
  contexts of, 11, 144, 154
Matrix
  concepts of, 55–56
McGreevy, Paul
  references to, 57
McLean, Andrew
  references to, xvii, 23, 65, 78–79, 85
McVey, Jones
  references to, 1, 55
Meaning, xiv–xv, xvii, 1, 7–10, 13, 14, 16, 26, 31, 35–38, 41, 54, 59, 69, 80, 82, 88, 98, 103, 108, 114, 118, 124, 131, 142, 146–147, 151, 166
Measureed/Measurements, 9, 27, 83, 100, 102, 121, 141, 155; *see also* Indicators; Metrics
Media
  contexts of, 10, 58–60, 105, 111, 121, 143
Medical/Medicine, 78, 90–92, 106
Mellor, David
  references to, 153, 158, 163
Memory
  concepts of, 86, 133
Mental/Mentality/Mentally, xvii, 2–5, 7, 9–15, 18, 22, 24, 27, 30, 32, 34–36, 42, 45, 48, 61–62, 64, 70–71, 77–80, 82, 87–88, 103–104, 114, 121–122, 126, 137, 145, 151, 153, 157–159, 162–163, 167; *see also* Affect; Cognition
Mid-information, *see* Midline; Misinformation
  anatomy and posture, 57–59
Mirrors, 25, 84, 105–107, 110, 123; *see also* Feedback; Reflection
Misguided
  concepts of, 58–59
Misinformation/Misinformed, 10, 56–58, 60; *see also* Communication; Infodemic; Media
Mistakes
  concepts of and examples, 39, 73, 120, 134–135, 138, 140–141; *see also* Error; Learning
Mitigate
  acts of mitigating, 13, 18, 79, 122
Mode
  contexts of, 6, 31, 73, 96
Models, 3–5, 10–11, 27, 29–30, 33, 40–46, 59–60, 62, 64, 66, 69, 81, 99, 109, 112, 135, 143, 146; *see also* First Principles; Five Domains
Modern
  contexts of, xvii, 10, 23, 65, 141, 154
Mood, 38, 44, 133
Moral
  concepts of, 10, 110, 116, 151; *see also* Ethics
Motivation, 4, 6–7, 10, 13, 27, 33, 34, 36–39, 43, 46–49, 53, 57, 59, 70, 133, 141, 145, 155; *see also* Reinforcement
Mounting, xii, xv, 24, 43–44, 64, 72–73, 92, 141, 156

Mouth
contexts of, xv, 4, 12, 27, 42, 49–61, 74, 152
references to, xv, 94, 156; *see also* Aids; Biting; Contact; Muscles

Mutual
concepts of, 11, 19, 25, 42, 80, 134–135

Narrative
concepts of, 8, 105, 112; *see also* Framing; Story

Nasal
references to, 45, 49, 155, 160, 164; *see also* Nose

National
contexts of, 32, 83, 110; *see also* Governance; Policy

Natural/Naturally/Nature, 8, 14–15, 20–21, 33–34, 51–52, 66, 93, 140; *see also* Behaviour; Environment

Naughty
uses of, xviii, 7–8, 34, 36, 59, 71, 85, 91; *see also* Labels; Misinterpretation

Navigating, 3, 32, 59, 61, 69, 70, 75, 86, 106, 118; *see also* Decision-making

Neck
anatomy and posture, 1, 25, 38–39, 71, 77, 109, 123, 155, 159–160, 164–165; *see also* Head; Posture

Negative
concepts of, xv, xvii, 3, 5, 7–8, 13–16, 18–19, 22–23, 28, 30–37, 44, 53, 61, 74, 78–79, 81, 88, 92, 107–109, 140–141, 145, 151, 154, 158, 163; *see also* Punishment; Reinforcement

Nervous
states of, xv, 122, 155, 159, 164

Networks
concepts of, 60, 97, 99, 143; *see also* Collaboration; Community

Neuroscience, 28–29, 33, 35, 43, 73, 85

Neurotransmitter, 28, 141; *see also* Brain; Learning

Neutral
neutral states and aids, 16, 18, 24, 157, 162, 167

Noises
sensory context, 17, 87; *see also* Arousal; Hearing

Non-equestrians
references to, 35, 42, 99, 112; *see also* Public

Norms, 11, 52, 69, 111; *see also* Culture; Practice

Nose, xviii, 49, 67; *see also* Gauge; Head; Restraint, 1, 11, 13, 26, 32, 45, 49–50, 61, 108–109, 111, 141, 148, 155, 160, 164

Novelty, 3, 9, 32, 81, 124, 160; *see also* Arousal; Enrichment

Objective, 14, 17, 32, 57, 61, 86, 122, 137, 154–155, 158, 159, 163, 164; *see also* Evidence; Evaluation

Observations, xvii, 19, 20, 22, 24, 45, 47, 51, 61–62, 70–81, 95, 97, 115, 122, 126, 133–140, 157, 162, 167; *see also* Awareness; Behaviour; Measurement

Olympic
references to, 12, 27, 90; *see also* Competition; Sport

Online
contexts of, 10, 12, 14, 26, 42–43, 51, 60, 66, 68–69, 79, 97, 100, 116, 132, 136, 144, 146; *see also* Media

Openness, 4, 26, 40, 48–49, 61, 82, 125, 126, 152, 158, 159, 161, 162, 164, 167
open-ended, 45–47, 147, 148; *see also* Curiosity; Learning environments
Operant
concepts of, 6, 49, 73–74, 154, 158, 163; *see also* Conditioning; Reinforcement
Opinions
concepts of, 47, 50–52, 97, 129, 137; *see also* Perspective; Feedback
Opportunity, 9, 13–20, 26–28, 33–36, 38–39, 42, 43, 48, 53, 64, 77, 79–81, 92, 100–101, 108–110, 123–125, 128, 129, 138–140, 142, 147, 151, 157; *see also* Agency; Possibility
Opposed/Opposing/Opposite, 4, 7, 12, 18, 28, 33, 41, 44, 58, 60, 71, 75, 77, 79, 87, 91, 115, 140, 145, 160–162, 165–167; *see also* Contrast; Polarity
Optimal, 27, 49–50, 73, 76, 82, 85–88, 125, 155–156, 158, 162–164, 167
Organisations, xii, 10, 31, 53, 64, 65, 86, 90, 97–99, 109, 132, 143, 147; *see also* Governance
Origins, 3, 90, 93, 100; *see also* Heritage; History
Othering, 41, 43, 60
Outcomes, xvi, xviii, 1–2, 4–5, 7, 13, 19, 22, 28–29, 32, 34, 39, 52, 72–74, 96–100, 103, 106, 111–112, 131, 142, 144, 151; *see also* Change; Evaluation; Evidence; Ripples; Outdated; Overstory

Owner
concepts of, 39, 46, 86, 108, 122, 126, 129; *see also* Guardians; Stewardship

Pace
concepts of, 3, 9, 13
Pain
concepts and contexts of, xvii–xviii, 1, 4, 12, 15, 18, 27, 38, 56, 58, 71, 79, 85, 91, 112, 128, 149, 152
Pairing
and dyads, xv, 13, 35, 122–123
Paper
references to, 9, 31, 94
Paradigm
concepts of, xvi, 3, 93, 103–104, 107
Parasympathetic
references to, 123, 155, 159, 164
Park
contexts of, 43, 76, 156
Patting
contexts of, 11, 18, 25, 59, 67
Peer, 10–12, 87–88, 147
-reviewed, 64, 91, 137
Perception
concepts of, 13, 86
Perform, xiv, 16, 86, 152
Performance/s, 6, 8–9, 14, 32, 68, 86, 104–105, 113, 133, 144
Perspective/s, xix, 3, 13, 31, 32, 40, 42, 53, 59–61, 69, 89, 104, 111, 148
Phenomenon
references to, 7, 42, 96, 98, 115, 145
Phone
references to, 23, 45, 54–55, 103, 139
Physical, xvii, 2, 6, 9, 13–15, 22, 24, 31, 32, 34, 38, 45, 55, 62, 67, 74, 77–78, 80, 82, 87–88, 97, 100, 104, 110, 114, 119, 121, 151,

157–159, 162–164, 167;
see also Biomechanics;
Health; Picked uses of,
92, 155
Plane
  contexts of, 45, 49, 124, 155,
      160, 164
Play/s, 105, 107, 152
  role, 97, 102
Poll
  references to, 77, 123
Poor
  contexts of, 21, 157, 162, 167
Position/lity/ed/s, xvii–xviii, 3,
      38, 40, 41, 53, 61, 77,
      92, 157, 162, 167; *see also*
      Alignment; Posture
Positive, xvii, xix, 1, 3, 8–10,
      13–20, 22–23, 26–32, 34,
      36, 38–39, 41–48, 50, 53,
      62, 64, 66, 68, 70, 72,
      74, 79–82, 84, 85, 89, 90,
      92, 97, 98, 104, 107–111,
      126, 130; *see also* Affect;
      Reinforcement
Pracademic
  references to, 33, 65
Practical
  contexts of, 43, 65, 99, 153, 158,
      163
Practice, xvii–xviii, 2, 4, 6, 9,
      12–13, 16, 22, 25–26, 32,
      57, 61–62, 64–65, 68,
      79, 87–89, 92, 98, 102,
      105, 106, 108–111, 116,
      119–120, 123–125, 130,
      133, 134, 139, 147–151,
      156, 162, 167
Practitioners, 23, 69, 90–91
Pragmatic
  contexts of, 43, 66, 78, 97, 104
Praxis
  concepts of, vii–ix, 2, 6, 13–14,
      24, 27, 29, 56, 66, 80, 84–
      85, 89, 105–108, 114–115,
      118, 121, 124, 129–131,
      138, 143, 148–151, 155

Prediction, xviii,
      11, 28–29, 64, 72–75, 77,
      103, 141, 154–159, 161,
      162, 164–167
Pressure, xvi, xix, 3, 4, 6–7, 10, 13,
      17, 23, 25–29, 33–38, 44–
      49, 53, 58, 60–61, 66–68,
      72–74, 77, 112, 141–142,
      145, 151–156, 158–167
Principles, xvii–xviii, 7, 12,
      49, 57, 61–62, 64–66,
      69–79, 81, 83–85, 87,
      92, 95–100, 102, 108,
      115–116, 122, 136–137,
      142–145, 147, 149–154,
      158, 163
Problems, 7–8, 21, 43, 46, 51, 53,
      54, 61, 65, 87, 94, 104,
      107–110, 125
Progress, xix, 8, 9, 32, 45, 54, 60,
      77, 83, 91, 111, 140, 148,
      150, 157, 161, 166
Propensity
  concepts of, 40, 89, 93,
      145–147
Protocols
  references to, 20, 57, 112
Proximity
  concepts of, 16–17, 40, 65
Psychics
  references to, 93–94
Psychologists, 86, 88, 94, 103
Psychology/ical/ically, 36, 38, 58,
      59, 74, 82, 85, 88, 93–94,
      96, 100, 103, 113, 115,
      133–136, 138, 143, 144
Public
  contexts of, xviii, 1, 5, 13,
      23–24, 35, 62, 87, 92,
      106, 107, 109, 112, 118,
      141, 148
Punishment
  references to, 26, 116

Quality, 9–10, 15–16, 21, 33, 34,
      80, 87, 99, 108, 132, 147

Quantities, 20–21
Questions, xvi, 14–15, 23–26, 33, 41, 45–47, 49, 51, 53, 58, 62, 64, 76, 85, 89, 92, 94, 95, 103, 106, 126, 134, 141, 147, 148, 164

Redesigning, 8, 81
Reflection/Reflective, xvi–xvii, xviii, 2, 10, 13, 19–20, 22, 24, 28, 31, 43, 46, 54, 89, 93, 104, 118, 134, 135, 137, 138, 148, 150
Reflexive/Reflexivity, vii–ix, 2, 6, 13–14, 22, 24, 27, 29, 32, 33, 43, 46, 56, 66, 75–76, 80, 84–85, 89, 92–93, 96, 102, 105–108, 114–115, 118, 120, 124, 129, 131, 138, 143, 148–151, 155
Reform, 1, 82–85, 88–90, 96, 98–99, 104, 111
Refusing, 109, 113, 129
Regret, 10
Reins, xiv, 4–7, 25, 27, 28, 35, 40, 45, 49–50, 58, 60–61, 68, 78, 79, 97, 98, 141, 145, 147, 152, 154, 161, 165
Reinforce/ed/ment/er/ing, xiv, 10, 11, 13, 18, 37, 45, 49, 60, 66–67, 69, 73–78, 97, 98, 122–123, 130, 137, 146–147, 154, 163, 165–167
Relate/Relation/Relational/Relationship/Relationships, 9, 19, 20, 24, 29, 33, 36, 41, 78, 81, 92, 111, 130, 132, 133, 138, 149
Relaxation/Relaxed, 17, 20, 44, 76–77, 123, 155, 157, 159, 162, 164, 167
Release, xiv, 6, 13, 25, 27–29, 49, 66, 72–74, 78, 115, 141, 146, 152–156, 158–161, 163, 166

Relentless, 4, 23, 26, 37, 44, 127, 141, 167
Removal, 18, 28–29, 37, 38, 50, 69, 73–74, 76, 123, 146–147, 154, 158, 163
Repetition, 73, 75, 146, 156, 161, 165
Research, xviii, 5, 9–12, 23, 28, 35, 38, 40, 43–44, 50, 53, 55–59, 61, 64–66, 70, 74, 85–87, 92, 102, 104, 105, 109, 113, 115, 121, 131–134, 139–146
Resilience/Resilient/Resiliency, xviii, 34–35, 45, 62, 64, 78–79, 105, 122–127, 151, 157
Resist/Resistance, xix, 3, 33–34, 38, 57, 61, 68, 90, 102, 135, 140–143, 145–149
Resolution, 18–22
Resolve, xv, 3, 5, 7–8, 16–19, 22, 30–31, 33, 36–37, 49, 78–79, 81, 92, 108, 152
Resources, 41, 46, 126, 146
Respect, 8, 71, 104–105, 116, 118, 135
Responses, 7, 16, 23–24, 27, 44, 62, 68, 77, 82, 86, 123, 130, 153–154, 157–159, 162–164, 167
Responsivity, 17, 55, 77, 146, 156, 165, 167
Rest, 18, 81, 122–123, 152
Rethink, 10, 27, 53
Reverse, 136
Reward/Rewarding, 34, 42, 59–60, 68, 97–99, 141, 152
Right/Righting, xiv, 4–5, 8, 10, 13, 24, 27, 32–37, 45, 46, 48, 51, 54, 61–62, 72–73, 84, 92, 102, 103, 106, 114, 128, 130, 144, 161
Ripples, 9–10, 44, 99, 125, 130, 140

Risks, 19, 22, 59, 66, 91, 97, 107, 110, 127, 152
Roadmap, 1, 5, 119

Saddle, 54, 55, 72
Safe/Safeguard/Safeguarding, xviii, 19, 26, 39, 51, 57, 61–62, 76–78, 96, 99, 109, 111, 113, 133–138, 151
Safety, 66, 68, 76–77, 134, 147, 154, 159
Scholar, 65
Scholarship, 65
Science, xvi–xviii, 1–2, 6–8, 10–11, 13, 15, 18, 20, 25, 33–35, 37, 40, 44–45, 52, 57, 65, 70, 73, 85–86, 88, 92–94, 97, 99, 111, 115–116, 126, 137, 146
Scratching, xv, 13, 73–74, 78, 122–123, 141, 145, 155, 156, 160–161, 164, 166
Self-carriage, 27, 29, 34, 67, 146, 154–157, 159, 162, 167
Self-deception, 85, 87–89, 93, 95–96
Self-leadership, 34, 85, 104, 120
Self-reflection, 26, 148
Self-regulation, 122–123
Sense, 3, 11, 19, 53, 61, 68–70, 75, 80–81, 84–86, 97, 126, 132–133, 137
Sensitivity, 156, 162, 167
Sensory, 16–17, 70, 123, 137
Sentience/sentient, 5, 8, 12, 41, 50, 52, 55, 83, 108
Shame/Shaming, 121, 126–128
Shaping, 72, 74, 77–78, 92, 98, 116, 138, 154, 159, 163
Shoulder, 161, 165
Signals, xvii–xviii, 4, 7, 28, 60, 66, 76–77, 99, 154, 159, 163
Signpost, 17, 43, 79, 132
Signs, 17, 19–21, 61, 66, 76, 120, 152
Skills, xviii, 5–6, 11, 23, 60, 77, 100, 102, 104, 127, 129, 132, 151, 158, 163
Smoothie, 116–117
Social behaviour, 2, 8, 11, 19–20, 34, 39–43, 53, 56, 59–61, 68, 70, 89, 97–99, 104, 111, 116, 118, 119, 127, 132–133, 143, 148
Society, xvii, 7, 9, 11, 15, 25, 68, 70, 84, 90, 108, 138
Sorting, 41–43, 60
Species/Species-specific, 8–9, 13–18, 25, 32, 53, 57, 60, 71–72, 76–78, 80, 123, 142, 143
Speed, 27, 66–67, 73, 116, 155, 159, 164
Spooking, 12, 72, 154, 159, 164
Sports, xvi, 1–5, 8, 11–13, 15–18, 25, 27, 35, 41, 43, 50–52, 59, 61, 64, 81–82, 87–90, 96–99, 104–110, 112–114, 116, 118, 120, 123, 137–146, 148–152
Stables, 16, 20, 22, 24, 37, 39, 41, 58, 70, 110, 115
Status, 9, 12, 69, 82, 108, 110, 136–138, 142–145, 150
Step/Stepping/Steps, xvi, 10, 20, 27, 37, 43, 45, 47, 55, 68, 71–73, 77, 99, 106, 120, 134, 151, 156, 159, 160, 165
Stimuli, 18, 76–77, 79, 124
Stop/Stopped, xvii, 4, 35, 39, 42–43, 47, 49, 52, 73, 77, 82, 153, 155–156, 160–161, 165–166
Stress, 20–21, 37, 66, 112, 123
Strides, 27, 38, 77
Suffering, 15, 34, 70–71, 81, 106–107, 109
Sunscreen, 18, 79
Support, 5, 13, 59, 66, 70, 99, 107, 111, 114, 120, 130, 134, 136

Sustain/sustainability/sustainable/sustaining, 9, 31, 43, 107, 109, 111, 112, 151, 155, 164
System/Systematic/Systemic/systems, xvii, 21–22, 32, 50, 54, 71, 81, 87, 88, 101, 103, 111, 114, 123, 133, 137, 142, 155, 159, 164

Tack, 18, 24, 26, 35, 37, 50, 74, 126, 149
Taper-gauge, 13, 155, 160, 164
Teach/Teaching, xiv, 65, 155, 158, 159, 163, 164
Team, 51, 92, 102, 105, 134, 135
Telos, 8–10, 13–15, 32, 34, 76, 114, 131
Tenet, 27, 29, 32, 97, 108, 142
Tension, 4, 5, 26, 33, 45, 47, 138, 152
Threats, 37, 68, 71, 72, 82, 118, 148
Threshold, 17, 77, 143–144, 149
Thrive, 8–9, 53, 57, 61, 69, 80–81, 83, 84, 109
Timing, 66, 69, 73, 154–156, 158–159, 161–164, 166–167
Tongues, 1, 5, 12, 60, 109, 111
Tools, 4, 29, 57, 58, 61, 112, 116, 118, 152
Touch, 17, 26, 39, 73, 110, 122, 123, 155, 159, 164–165
Tradition/Traditional, xvii, 1, 2, 5–6, 8, 9, 11, 13, 56–60, 69, 87–90, 97, 135, 137, 151
Train/Trained/Trainers/Training, xvi–xvii, 3, 6–7, 10, 12, 17, 24–26, 28, 29, 33, 37–39, 43–45, 48, 51–55, 57, 64, 72–78, 83–85, 87–90, 97–98, 100, 102, 104, 108, 111, 115–116, 122–123, 125, 126, 137, 141, 146, 149, 153–167
Transforma, xix, 58, 96–97, 104, 112, 142, 144
Treating, 49, 50
Trials, 86, 91, 165
Triple, 9, 70, 110
Turns, 6, 35, 37, 41, 48, 71, 77, 82, 88, 91, 158–163, 167
Tversky, 103–104

Unambiguous, 154, 159, 163
Uncertain/Uncertainty, 3, 6, 10–11, 51–52, 60, 82, 110, 127
Uncomfortable, 3, 38, 54
Under-saddle, 76, 77, 123, 155, 156, 158, 159, 161, 163–166
Unlearn, xviii, 5–6, 10, 12–13, 15, 23, 25–27, 36, 145, 150–151
Unravelling, 82, 109
Unsafe, 43, 109, 134, 135, 138
Unwanted, 4, 17, 36, 46, 65, 86, 91, 102, 116, 122
Update/Updated, xvi–xix, 2–4, 9, 13, 15, 23, 26, 28–29, 39–40, 42, 48, 54, 56, 60, 74, 102, 115, 125, 142, 146, 147, 151
Upstand/Upstanding, 85, 106–107, 162, 167

Valence, 3, 79, 154, 157–158, 162, 167
Valuable, 105, 122, 139, 147
Value, ix, xvi, xix, 16, 23, 36, 41, 42, 50–51, 53, 59, 67, 69, 77, 96, 99, 102, 105–107, 114, 118–121, 125, 127, 132, 137, 140, 147, 149, 150, 157, 162, 167
Vault, 116
Verbal, 35, 73–74, 86, 122, 128
Veterinary, 18, 65, 102, 122–123
Vets, 12, 102
Vulnerability/Vulnerable, 60, 112–114, 127, 128, 136, 138, 140

Waran, 110, 111, 121
Warm-up, 155, 160, 164
Weaving, 21, 81
Welfare, 1–20, 22–23, 29, 31–35, 37, 39, 40, 43–45, 47, 48, 50–53, 55, 59–60, 62, 64, 68–72, 76, 79–90, 93, 96–97, 99, 100, 102, 104–121, 123, 131, 132, 134, 137–153, 157, 158, 163
 -centred, 25, 27, 65
 -focused, 26, 61, 98, 149
Well, 5, 13, 16, 22, 31, 59, 80–82, 106, 109, 123, 130, 143, 147
 -being, 3, 14, 45, 64, 66, 80–84, 88, 96, 102, 104, 105, 108–111, 114, 121, 131, 148, 150, 158, 163
Wheel, 25, 100–101, 104

Whip/Whipping, xviii, 1, 51, 72, 108, 111, 113, 116, 141, 160, 165–166
Whole, 2, 11, 42, 79, 103, 106, 119, 133–135
Wilkins, 30
Winning, 12, 23, 51, 87, 114, 138
Wither, 11, 13, 18, 25, 45, 59, 123, 141, 155, 159, 160, 164
Witness/Witnessed/Witnessing, 1, 39, 43, 50, 73, 75, 76, 77, 106, 157, 162, 167
Wrong, 10, 23, 48, 51, 69, 94, 114, 128–129, 138

Yards, 41, 43, 45, 65, 68, 75, 112, 115, 136, 138, 158, 163
Yield, 72, 77, 162, 165–166
YouTube, 102, 131, 142, 144

Zoos, 8–9

ISES Training Principles
Human and horse welfare depend upon
training methods and management that demonstrate:

1. Regard for human and horse safety
By acknowledging the horse's size, power and flightiness | By learning to recognise flight/fight/freeze behaviours early.
By minimising the risk of causing pain, distress or injury | By ensuring horses and humans are appropriately matched.

2. Regard for the nature of horses
By meeting horse welfare needs such as foraging, freedom and equine company | By respecting the social nature of horses.
By acknowledging that horses may perceive human movements as threatening | By avoiding dominance roles during interactions.

3. Regard for horses' mental and sensory abilities
By acknowledging that horses think, see and hear differently from humans | By keeping the length of training sessions to a minimum.
By not overestimating the horse's mental abilities | By not underestimating the horse's mental abilities.

4. Regard for emotional states
By understanding that horses are sentient beings capable of suffering | By encouraging positive emotional states | By acknowledging
that consistency makes horses optimistic for further training outcomes | By avoiding pain, discomfort and/or triggering fear.

5. Correct use of desensitisation methods
By learning to apply correctly systematic desensitisation, over-shadowing, counter-conditioning and differential reinforcement.
By avoiding flooding (forcing the horse to endure aversive stimuli).

6. Correct use of operant conditioning
By understanding that horses will repeat or avoid behaviours according to their consequences | By removing pressures at the onset
of a desired response | By minimising delays in reinforcement | By using combined reinforcement | By avoiding punishment.

7. Correct use of classical conditioning
By acknowledging that horses readily form associations between stimuli.
By always using a light signal before a pressure-release sequence.

8. Correct use of shaping
By breaking down training into the smallest achievable steps and progressively reinforcing each step toward the desired behaviour.
By changing the context (trainer, place, signal), one aspect at a time | By planning the training to make it obvious and easy.

9. Correct use of signals or cues
By ensuring the horse can discriminate one signal from another | By ensuring each signal only has one meaning
By timing the signals with limb biomechanics | By avoiding the use of more than one signal at the same time.

10. Regard for self-carriage
By training the horse to maintain gait, tempo, stride length, direction, head, neck and body posture.
By avoiding forcing a posture or maintaining it through relentless signalling (nagging).

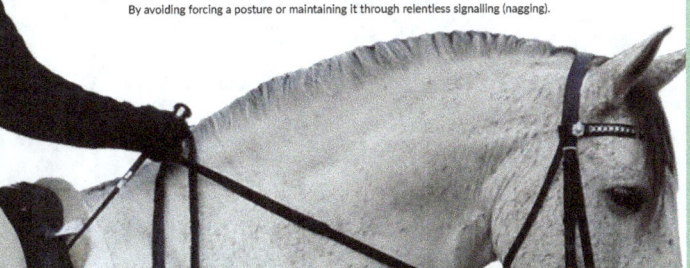

This poster is a summary of ISES Training Principles. To read the extended version go to:
www.equitationscience.com

For Product Safety Concerns and Information please contact our EU
representative GPSR@taylorandfrancis.com
Taylor & Francis Verlag GmbH, Kaufingerstraße 24, 80331 München, Germany

www.ingramcontent.com/pod-product-compliance
Lightning Source LLC
Chambersburg PA
CBHW060609230426
43670CB00011B/2040